Generis
PUBLISHING

AF510381

Sweet potato (*Ipomoea batatas* Lam): a rich source of bioactive components and a promising crop for health promotion

Prof. Predner Duvivier

Title: **Sweet potato (*Ipomoea batatas* Lam): a rich source of bioactive components and a promising crop for health promotion**

ISBN: 979-8-89248-918-8

Author: Prof. Predner Duvivier

Cover image: https://pixabay.com/

Publisher: Generis Publishing
Online orders: www.generis-publishing.com
Contact email: info@generis-publishing.com

中文摘要

學號: P9522007

論文題目:甘藷（*Ipomoea batatas* Lam.）之抗氧化研究

總頁數: 307

學校名稱:國立屏東科技大學

系（所）別:熱帶農業暨國際合作系

畢業時間及摘要別: 九十七學年度第二學期博士學位論文摘要

研究生: 杜維耶

指導教授:謝寶全 博士

賴博永博士

摘要論文內容

全世界甘藷塊根的年產量超過 133 億噸，分別為第五及第七大宗的主要糧食作物。本研究評估並比較三種不同品系的甘藷（*Ipomoea batatas* Lam.）:台農 57、台農 66 及台農 73 之抗氧化活性、酚類及類胡蘿蔔素含量與顏色參數，研究主要探討因子有萃取溶劑、不同甘藷品系、不同甘藷部位、乾燥方式、分離方法、貯存溫度及貯存時間的影響。

結果顯示萃取的溶劑中:以乙醇和乙醚萃取所得的平均總酚含量較高，之後依序為丙酮、甲醇、乙酸乙酯、50%乙醇及去離子水；以 DPPH 法評估抗氧化活性，結果顯示以甲醇、乙醇及丙酮的萃取物平均抗氧化活性較高，依序為乙醚、乙醇水溶液、乙酸乙酯及去離子水。由於以乙醇萃取具有較高的抗氧化效果且相對於丙酮、乙酸乙酯及甲醇，其毒性較低，因此選為進一步之萃取溶劑。在不同品系間，其多酚含量及抗氧化活性含量依序為台農 73＞台農 66＞台農 57；台農 66 品系則表現出較高含量之類胡蘿蔔素，而台農 57 及台農 73 中的總胡蘿蔔素含量極低，甘藷皮之抗氧化活性及酚含量均高於果肉的部位。

以冷凍乾燥處理的樣品其抗氧化活性及酚類含量平均高於以空氣乾燥處理的樣品。空氣乾燥法包含有日曬、低溫乾燥（25°C）及熱風乾燥（50ºC 及 75ºC）共 5 種，結果顯示以 DPPH、linoleic acid 抗氧化系統（AOP）和 ABTS 測定後三種不同品系之甘藷其抗氧化活性均明顯下降。抗氧化活性的下降歸因於總酚及總類黃酮含量的下降。而其中台農 73 之抗氧化活性，在熱風乾燥過程，抗氧化活性、總酚及總類黃酮含量會隨溫度增加而下降。然而台農 57 及台農 66 的抗氧化活性、總酚及總類黃酮含量會隨著乾燥溫度增加及時間減少而增加。由結果顯示食品在乾燥過程中，抗氧化物質及其活性的變化和品系具有相關性，低溫乾燥法不但費時且製程難以放大，而台農 73 品系，較適合以乾燥處理。

由於此三種抗氧化分析方法之間缺乏相關性，因此利用決定係數（R^2）及變異係數（CV）來探討它們之間的最適配合度；結果顯示，這些方法之間因缺乏足夠的證據來指出哪個方法較佔優勢，因此利用單一抗氧化活性指標（single AOA indicator），也稱作抗氧化相關的期望指標（relative desirability index），來評估 DPPH、AOP 及 ABTS 之能力，以克服不同方法間缺乏相關性的問題。台農 57 品系，經冷凍乾燥後其 RDI 值最高，日曬乾燥最低，相較於多酚及胡蘿蔔素含量，以胡蘿蔔素含量與 RDI 值有較高的相關性；而台農 66 品系，經冷凍乾燥後其 RDI 值最高，日曬乾燥最低。在低溫與熱風乾燥（50 ºC 及 75 ºC）時，台農 57 及台農 66 品系，其 RDI 值會伴隨乾燥溫度增加而降低；台農 73 品系，以冷凍乾燥法處理其 RDI 值最高，其次為 75°C 熱風乾燥、日曬乾燥及 50°C 熱風乾燥，以 25 °C 處理的樣品其 RDI 值最低。於台農 66 及台農 73 品系，其多酚含量較胡蘿蔔素含量呈現較高的 RDI 相關性。不同品系之最適乾燥條件，分別為台農 73 品系: 75 ºC，24 小時、台農 57 品系及台農 66 品系，為 25ºC， 10 天。

於台農 73 品系中含有三種主要的酚類化合物；亦即，兒茶素（catechin）、芸香素（rutin）及沒食子酸（gallic acid），其在萃取物中的含量依序為 268.24 mg/kg、67.06 mg/kg 及 25.56 mg/kg。在甘藷不同部位方面，皮的部份以 catechin 及 gallic acid 含量較高，rutin 含量少於果肉的部份，在台農 73 中主要的抗氧化活性來源為 rutin，依序為 catechin 及 gallic acid。

在貯存試驗方面，樣品貯存在 4℃及室溫下其總酚含量並沒有顯著差異，但抗氧化活性以貯存在 4℃較室溫為佳；然而，總酚含量和抗氧化活性，均會隨貯存時間增加而下降；甘藷果肉經貯存在室溫 9 個月及 4 ℃貯存 10 個月其總酚含量會顯著下降，而甘藷皮經貯存在 4℃及室溫，11 個月總酚含量；甘藷果肉貯存在室溫，5 個月及 4 ℃，10 個月其抗氧化活性均會顯著下降，而甘藷皮經貯存於 4℃，11 個月其抗氧化活性會顯著下降。

糊化分析方面，結果顯示三種不同品種的尖峰黏度及最終黏度，分別超過 495 及 335 RVU，且尖峰黏度在 80 ℃，5 分鐘內，由以上結果顯示，三種不同品種的甘藷其萃取殘留物可以應用在原料需要適度高糊化的凝膠食品或其它食品，由本研究的結果中證明將地瓜萃取後的殘留物製成布丁、餅乾及蛋糕是可行的。

關鍵字: 甘藷、抗氧化活性、化學組成、乾燥方法、相對期望指標

Abstract

Student ID: P9522007

Title of Dissertation: Antioxidant activity of sweet potato (*Ipomoea batatas* Lam.)

Total Pages: 307

Name of School: National Pingtung University of Science and Technology

Name of Department: Department of Tropical Agriculture and International Cooperation

Graduate Date: June 29, 2009

Degree Conferred: Doctor of philosophy (Ph. D.)

Name of Student: Pedner Duvivier

Advisors: Pao-Chuan Hsieh (Ph. D.) and Po-Yung Lai (Ph. D.)

The Contents of the Abstract in this Dissertation:

Sweet potato production and utilization exceeds 133 million tons of tuberous roots annually in the world. It represents the 5th and 7th staple food crop in developing countries and the world, respectively. This study evaluated and compared three sweet potato (*Ipomoea batatas* Lam.) cultivars: Tainong 57 (TNG57), Tainong 66 (TNG66), and Tainong 73 (TNG73) on the basis of their antioxidant activity (AOA), phenolic and carotenoid contents, and color parameters. The effects of extracting solvent, genotype, part of the root, drying method, fractionation method, storage temperature, and storage duration were tested.

The results showed that on average, ethanol and diethyl ether (DE) extracts exhibited the highest total phenolic contents, followed in descending order by acetone, methanol, ethyl acetate (EA), aqueous ethanol (1:1 v/v) and distilled de-ionized water (ddH$_2$O) extracts. The highest AOA, as evaluated by 1, 1-diphenyl-2-picrylhydrazyl (DPPH) assay, were observed in methanol, ethanol, and acetone extracts, followed in descending order by DE, aqueous ethanol, and EA, and ddH$_2$O extracts. Due to its high effectiveness and its relative lower toxicity compared to acetone, DE and methanol, ethanol was chosen for further extractions.

Among the three cultivars, TNG73 had on average the highest phenolic content and AOA, followed by TNG66 and TNG57. TNG66 exhibited the highest carotenoid content, while TNG57 and TNG73 showed low content of total carotenoids. The skin samples had, on average, a higher AOA and phenolic content than the flesh samples.

Freeze-dried samples had on average higher AOA and phenolic content than the air-dried samples. During the air-drying process, including sun-drying, low temperature air drying (LTD) at 25°C, hot air drying (HAD) at 50°C, and HAD at 75°C, we observed in the three cultivars significant decreases of the AOA, as evaluated by DPPH, anti-oxidative potency in linoleic acid system model (AOP); and 2,2'-Azino-di-[3-ethylbenzothiazoline-6-sulfonate] (ABTS) assays. The decrease of AOA was attributed to a decrease of total phenolics and total flavonoids. The losses of AOA, total phenolics, and total flavonoids decreased with the increase of the drying temperature and the decrease of the drying time of TNG73 during the air-drying process. However, in TNG57 and TNG66, an increase of the losses of AOA, total phenolics, and total flavonoids during the air-drying process was observed with the increase of the drying temperature associated with a decrease of the drying time. These results indicated that changes in antioxidant components and activities during food drying are genotype-dependant, as the trend of variation was different for the three cultivars. As LTD is time consuming and difficult to scale up, TNG73 was therefore the most suitable for being processed by drying.

Low correlations were observed among the three AOA analytical methods. Therefore, their relative goodness of fit was studied using the coefficients of determination (R2) and variation (CV). Based on the results, there was no sufficient evidence to show that one method was better than another. Thus, a single AOA indicator, referred to as AOA relative desirability index (RDI), was computed from the results of DPPH, AOP, and ABTS assays to overcome the problem of lack of correlation among the analytical methods used. For TNG57, the highest RDI value was observed in freeze-dried samples, while the lowest was observed in those treated by sun-drying. Higher correlation of the RDI was observed with carotenoid than with phenolic contents of TNG57. For TNG66, freeze-dried samples showed the highest RDI, followed by sun-dried ones. In LTD, HAD at 50ºC, and HAD at 75°C, the RDI decreased with the increase of the drying temperature for both TNG57 and TNG66. For TNG73, the highest RDI was observed in freeze-drying, followed by HAD at 75°C, then sun-drying, and then HAD at 50°C. The samples dried at 25ºC showed the lowest RDI. Both for TNG66 and TNG73, higher correlation of the RDI was observed with phenolic than carotenoid contents. The best air-drying conditions were at 75ºC for 24 h for TNG73, and at 25°C for 10 days for TNG57 and TNG66.

Three major phenolic compounds, namely, catechin, rutin, and gallic acid were identified in TNG73. Catechin (268.24 mg/kg dry matter) was the highest, followed by rutin (67.06 mg/kg dry matter), and then gallic acid (25.56 mg/kg dry matter), respectively. Comparing with the flesh samples, the skin exhibited higher contents of catechin and gallic acid, but had lower content of rutin. Rutin was the major contributor to the AOA of TNG73, followed by catechin, and gallic acid.

In a storage test, the samples stored at room temperature and 4ºC did not show significant difference in the total phenolic content, but the AOA was on average lower at room temperature than at 4ºC. Both the AOA and total phenolics exhibited a general trend of decrease during the storage. The decrease of the total phenolics of the flesh samples was significant after nine-month storage at room temperature and 10 months at 4ºC. For the skin samples, the decrease of total phenolics was significant after 11-month storage both at room temperature and 4ºC. The decrease of the AOA of the flesh samples was significant after five-month storage at room temperature and 10-moonth storage at 4ºC. For the skin samples, the decrease of AOA was significant after 11-month storage both at room temperature and at 4ºC.

Results of a pasting profile analysis showed that the peak and final viscosity of the starches isolated from all three cultivars exceeded 495 and 335 RVU, respectively, with a peak viscosity time shorter than 5 min around 80ºC. These results showed that the solid extraction residue of the three cultivars could be processed as gel foods and other food items that require moderate to high gelatinization ability of the raw material. Sweet potato-based puddings, cookies, and cakes prepared in this study supported our findings of the suitability and usefulness of the starches contained in the residue of sweet potato after extraction of antioxidants.

Keywords: sweet potato, antioxidant activity, chemical compositions, drying methods, relative desirability index

Acknowledgments

First of all, I want to express my deepest thanks and gratitude to my God who always makes a way when there seems to be no way.

I am thankful and grateful to:

§ My advisor (Dr. Pao-Chuan **HSIEH**), my co-advisor (Dr. Po-Yung **LAI**), and Dr. Albert L. **CHARLES** for their academic, logistic, and moral supports;

§ The committee members of my dissertation (Dr. Robin Y. Y. Chiou, Dr. Jeng-Leun Mau, Dr. Shyang-Chwen Sheu, and Dr. Albert Linton Charles) for their valuable reviews, comments, and suggestions;

§ My sponsor, Taiwan International Cooperation Development Fund (**ICDF**) for its financial support;

§ The **DTAIC** office staff for their professional and friendly administrative handling;

§ All the teaching staff of **DTAIC** and **NPUST** for their contribution to my education;

§ All my lab-mates, in particular Ms. Chen Ya-Hui, for their friendliness;

§ The Government of Republic of China (Taiwan) for its hospitality; and

§ The Government of Republic of Haiti for the opportunity.

Last, but not least, I would like to thank all the members of my family for their love, prayers, and moral support.

Table of Contents

List of Figures

List of Tables

PART I. GENERAL FRAMEWORK

Chapter 1. General introduction

1.1. Background information and justifications

Reactive oxygen species (ROS); such as superoxide (O^{2-}), peroxyl (O_2^{2-}), oxide (O^{2-}), hydroperoxyl, hydroxyl (OH^-), and alkoxyl; are free radicals that are often generated as byproducts of biological reactions (Lo *et al.*, 2004). They react with other molecules, starting a chain of oxidation reactions with damaging effects on the cells. The human body of a healthy individual also produces some antioxidants; such as superoxide dismutase (SOD), glutathione peroxidase, and glutathione reductase; which can counteract the damaging effects of free radicals in tissues by keeping them in balance (Rimbach *et al.*, 2005). Rimbach *et al.* (2005) also reported that other nutrients such as β-carotene (a vitamin A precursor), vitamin C, vitamin E and a microelement, selenium, have been found to act as antioxidants.

In case of imbalance of free radicals, that is, when there are few antioxidants in the organism to counteract them, significant damages can occur, playing a role in the aging process, cancer, coronary heart disease, and Alzheimer's disease (Grey, 1990; Diaz *et al.*, 1997). Free radicals have been also reported to ease the development of neurodegenerative disorders, atherosclerosis, cataracts, inflammation, malaria, and acquired immuno-deficiency syndrome (Halliwell and Gutteridge, 1989; Slater, 1989; Jenner, 1991; Riemersma *et al.*, 1991; Sies, 1991; Aruoma, 1993; Yoshikawa, 1993; Aruoma, 1994). Studies have suggested that antioxidants occur naturally in fresh fruit such as mango, papaya, and citrus; vegetables; and whole grains, and eating a diet rich in such antioxidant-containing foods, has been linked to a lower risk of the above mentioned diseases (Mathews-Roth, 1985; Bendich and Olson 1989; Bendich, 1990; Krinsky, 1990; Gerster, 1991; Mathews-Roth, 1991; Ziegler 1991; Byers and Perry, 1992; Krinsky, 1994; Tonuci *et al.*, 1995).

According to the report of Food and Agriculture Organization and World Health Organization (FAO and WHO, 2002), the burden of chronic diseases linked to inappropriate diet is rapidly increasing worldwide. It has been calculated that, in 2001, chronic diseases contributed approximately 60% of the 56.5 million total reported deaths in the world and 46% of the global burden of disease. The proportion of the burden of non-communicable diseases is expected to increase to 57% by 2020. Almost half of the total chronic disease deaths are attributable to cardiovascular diseases. Obesity and diabetes are also showing worrisome trends, not only because they already affect a large proportion of the population, but also because they have started to appear earlier in life. The chronic disease problem, traditionally limited to the developed regions of the world, is increasingly affecting populations in developing countries. In five out of six regions of WHO (Africa, America, Europe, Eastern Mediterranean, South-East Asian, and Western Pacific), deaths caused by chronic diseases dominate the mortality statistics. It has been projected that, by 2020, chronic diseases will account for almost three-quarters of all deaths worldwide.

To ensure food to be safe and nutritious, harmful effects of lipid oxidation should be prevented. In food industry, addition of free radical scavenging antioxidants has been one of the effective ways to retard fat oxidation of food. Synthetic antioxidants such as butylated hydroxyanisole (BHA), butylated hydroxytoluene (BHT), *n*-propyl gallate (PG), monoglyceride citrate (MGC) and *tert*-butylhydroquinone (TBHQ) are widely used (Li *et al.*, 2006). Nevertheless, consumers' demands for natural food ingredients have continuously increased, and there has been a general desire to replace synthetic food additives with natural options (Loliger, 1991; Block *et al.*, 1992; Shahidi and Wanasundara, 1992; Stavric *et al.,* 1992; Brownridge, 1993; Hertog *et al.,* 1993; Aruoma, 1994; Bonorden and Pariza, 1994; Laranjinha *et al.*, 1994). That is why plant-derived antioxidants such as vitamin E, vitamin C and flavonoids have been increasingly suggested as important dietary antioxidant (AOA) factors.

Therefore, the evaluation of food crops for AOA has become a big issue and has drawn the attention of the scientific community. As a contribution to the knowledge of dietary AOA, we analyzed AOA of three cultivars of sweet potato (*Ipomoea batatas* Lam.), namely, Tainong 57 (TNG57), Tainong 66 (TNG66), and Tainong 73 (TNG73). We also studied the distribution pattern of AOA between the skin and the flesh of the tuberous root.

Fogliano *et al.* (1999) reported differences in the results of AOA of food commodities extracted with different solvents and assessed by different methods. Thus, it is necessary to evaluate different extracting solvents and various analytical methods for determining AOA in the different plant parts of the three cultivars of sweet potato. In this study, we tested seven extracting solvents, including acetone, aqueous ethanol (1:1, v/v), distilled de-ionized water (ddH$_2$O), diethyl ether (DE), ethanol, ethyl acetate (EA), and methanol. AOA was assayed using 1, 1-diphenyl-2-picrylhydrazyl (DPPH), 2, 2-azino-di-[3-ethylbenzthiazoline-6-sulfonate] (ABTS), and anti-oxidative potency in linoleic acid system model (AOP) assays. The total phenolic and carotenoid contents of the skin and the flesh of the three sweet potato cultivars were also analyzed.

Food insecurity is still widely present in the world (FAO, 2006a). Sweet potato is an easy-to-grow crop with good adaptability to diverse environmental conditions, high yielding ability, and high energy content (Ravindran *et al.,* 1995). It is rich in dietary fiber, minerals, and vitamins (Woolfe, 1993) and can be grown with no synthetic pesticides and fertilizers (FAO, 1984). It is suitable for environmentally friendly agriculture and organic food production and is a promising crop for global food security (Scott, 1992). Sweet potato can be successfully cultivated everywhere throughout tropical and warm temperate regions where there is sufficient water to support its growth.

Sweet potato is currently cultivated in more than 100 countries, covering more than nine million hectares. Leaves and roots of sweet potato are edible. With its high nutritional values, sweet potato can play a role in cutting malnourishment of the people in developing countries. It ranks fifth most important food crop in terms of fresh matter after rice, wheat, maize, and cassava in developing countries where 95% of its production (estimated at more than 133 million tons annually) is concentrated (Internacional Center of Potato, 1998). Some cultivars have anti-carcinogenic properties and/or AOA (Teow *et al.*, 2007). However, neither all food crops, nor all parts of a food crop have equal AOA (Rodriguez-Amaya, 2003). With the flesh color varying from white, yellow to orange and even red among the cultivars, the AOA of sweet potato roots may vary over a wide range (Teow *et al.*, 2007). The AOA also varies with the post harvest process to which the food material is subjected.

Food drying and food storage are two important factors of food security (FAO, 2003). Food drying brings the food water activity to a level unfavorable for the microorganisms' development during storage and facilitates food distribution by reducing bulk volume. Storage guarantees food availability between two harvesting periods. Nevertheless, processing and lipid autoxidation via free radical chain reaction during the storage lowers the nutritional value of foods, deteriorates their flavor and taste, and may be a cause of numerous diseases (Aruoma and Halliwell, 1991).

Although drying is important to lengthen the shelf-life of foods and foodstuffs, it can destroy the secondary metabolites responsible for AOA if it is not conducted under proper conditions (Rodriguez-Amaya, 2001). Thus, the suitability of a drying process, like any post harvest food treatment, depends on its effect on nutritious and functional properties of the food (Marc *et al.*, 2004). Hence, in this study, we evaluated the effects of five drying processes, namely, freeze-drying, sun-drying, low-temperature air drying at 25°C (LTD), hot air drying (HAD) at 50°C, and HAD at 75°C on the AOA and phenolic and carotenoid contents of three test cultivars of sweet potato, namely, TNG57, TNG66, and TNG73. The evolution of the AOA and phenolic content of dried and milled sweet potato as a function of the storage temperature (4°C and room temperature) and duration (1 to 12 months) was also evaluated.

1.2. Objectives

1.2.1. Overall objective

The overall objective of this study was to evaluate the total AOA of sweet potato under the influence of genotype, extracting solvent, drying method, and storage conditions and to enhance a better awareness of the health benefit of this crop and its role in the food and nutrition security system in the world.

1.2.2. Specific objectives

More specifically, this study pursued to:

1. Evaluate the chemical compsition and AOA of three cultivars of sweet potato, namely, TNG57, TNG66, and TNG73.

2. Test the effects of seven solvents (acetone, aqueous ethanol (1:1, v/v), ddH_2O, DE, EA, ethanol, and methanol) on the AOA and total phenolic content of in the extracts of TNG57, TNG66, and TNG73 sweet potato.

3. Establish the distribution pattern of the AOA between the skin and the flesh of TNG57, TNG66, and TNG73 sweet potato.

4. Evaluate the effects of five drying methods (freeze-drying, sun-drying, LTD, HAD at 50°C, and HAD at 75°C) on the AOA of TNG57, TNG66, and TNG73 sweet potato cultivars.

6. Analyze the relative goodness of fit of three methods (DPPH, ABTS, and AOP assays) to measure the total AOA of TNG57, TNG66, and TNG73 sweet potato.

7. Study the phenolic profile of TNG73 sweet potato cultivar and the contribution of individual phenolic compounds to its AOA.

8. Study the evolution of the total AOA and phenolic content of dried and milled TNG73 sweet potato during 12-month storage as affected by the storage temperature (room temperature and 4°C).

9. Analyze the pasting profiles of the flours and starches isolated from TNG57, TNG66, and TNG73 sweet potato cultivars to determine the suitability and usefulness of the solid residue of sweet potato after antioxidant extraction to develop new food items.

10. Evaluate the possibility of storage sweet potato processed into flour to ensure availability of good quality food during inter-harvest seasons.

1.3. Hypothesis

Phenolic compounds, which are present in fruits and vegetables, have been known as strong antioxidant substances (Awad *et al.*, 2001; Ehlenfeldt and Prior, 2001). According to Kurata *et al.* (2006), sweet potato leaves have a high content of phenolics responsible for its anticancer properties. Teow *et al.* (2007) reported that the root also contains phenolics compounds. Their concentrations vary among cultivars and among different parts (Ooyanagui, 1984). Therefore, their AOA may also vary. In general, fruit peel contains a higher concentration of antioxidant substances in comparison to the flesh (Kondo *et al.*, 2002). It has been under the assumption that the same phenomenon can occur in sweet potato.

The AOA of fruits and vegetables is affected by the drying process to which the food items are submitted (Yen *et al.*, 1993; Prior *et al.*, 1998 ; Wilhelmina *et al.*, 1999; Tsai *et al.*, 2005; Heinäaho *et al.*, 2006; Rembiałkowska, 2006; Schulzová *et al.*, 2006; Yan, 2006). On basis of those findings, it was assumed that AOA in sweet potato varies among the cultivars and at different parts of a sweet potato plant, skin and flesh, and with the drying method and conditions used. The effects of the drying process vary with the genotype and the drying method used.

Aruoma and Halliwell (1991) reported that lipid autoxidation via free radical chain reaction during the storage lowers the nutritional value of foods, deteriorates their flavor and taste, and may be a cause of numerous diseases. Thus, the AOA decreases during the storage and the decrease rate varies with the storage time and temperature.

1.4. Research questions

This study was conducted to seek answers for seven research questions.

1. Among the sweet potato cultivars TNG57, TNG66, and TNG73, which is most suitable for use as functional food on the basis of its AOA?

2. Does the skin have higher AOA than the flesh as it was found in some fruits?

3. Which is the best solvent for extracting sweet potato antioxidant among acetone, aqueous ethanol (1:1, v/v), ddH_2O, DE, ethanol, EA, and methanol?

4. What is the best combination of temperature and time to maintain the AOA of sweet potato during the drying process?

5. Up to how long is the AOA of dried and milled sweet potato retained in storage and what is the role of storage temperature in affecting the level of the AOA?

6. How good are the pair-wise correlations among the analytical methods of DPPH, ABTS, and AOP assays used to measure the total AOA in sweet potato? In case of poor correlations, what is the soundest assay?

7. In case of insufficient evidence of superiority of one analytical method over the others, what can be done to solve the problem of lack of correlation?

PART II. LITERATURE REVIEW

Chapter 2. An overview of sweet potato, food constituents, and food analysis

Abstract

An overview of sweet potato was presented in this paper with an emphasis on its role in the food security system, its genotypic diversity, and its potentiality as a functional food for disease prevention and health promotion. A concise description of food constituents and analysis was also presented. More than 133 million tons of sweet potato were produced and utilized annually in the world; 95% of which was produced in developing countries. This amount of production ranked sweet potato in the 5th and 7th most important staple food in developing countries and the world, respectively. With more than 5000 genotypes collected, it shows that sweet potato (*Ipomoea batatas* Lam.) has been well collected, bred, grown or used for food and in agriculture in the world. However, sweet potato carried a stigma of "food for the poor", which has hampered its development and consumption. Some genotypes have been reported to possess antioxidant and/or anticancer properties. These reports suggested that it is worth an effort to continue the genotype evaluation and the improvement of processing and storage technologies for enhancing sweet potato usage to ensure global food security.

Key words: antioxidant activity, food constituents, functional food, genotypic diversity, sweet potato

2.1. An overview of sweet potato

2.1.1. General description

Sweet potato (*Ipomoea batatas* Lam.) is a crop plant belonging to the family of Convolvulaceae (Wanda, 1995), is an herbaceous perennial vine, bearing alternate heart-shaped or palmate lobed leaves and medium-sized sympetalous flowers. Sweet potato cultivars exist in many colors. The edible tuberous root is long and tapered in shape, with a smooth skin whose color may be red, purple, yellow, brown, or white. Its flesh may be white, yellow, orange, or red although white and yellow-orange fleshes are the most common.

2.1.2. Origin, distribution, and production

Sweet potato is native to the tropical Americas and was domesticated there at least 5000 years ago (Zhang *et al.*, 2000). Other works, based on morphological variation in *I. batatas*, indicated maximum diversity in the area comprising Colombia, Ecuador and northern Peru (Yen, 1982). It was already widely established in the Americas by the time Europeans first arrived there (Zhang *et al.*, 2000). Sweet potato spread through Asia and Africa during the 17th and 18th centuries. It was also known before westerners conducted exploration in Polynesia. It is suspected that a prehistoric dispersal of sweet potato ocurred, basing on the parallelled lexical on sweet potato between the Quechua word 'kumar' and the Polynesian word 'kumara'. The parallelled lexical on sweet potato indicates the possible human transfer of sweet potato from the neo-Tropics to eastern Polynesia (O'Brien, 1972; Yen, 1982).

Currently, sweet potato is cultivated throughout tropical and warm temperate regions wherever there is sufficient water to support its growth. The Consultative group on international agricultural research (CGIAR, 2006) reported that in 2004, approximately 129 536 275 tons of sweet potatoes were produced in more than 100 countries. According to CIP (1998), the annual production of sweet potato exceeds 133 million tons. Over 95% of the global production is located in developing countries where it ranks 5th most important food crop in terms of fresh matter after rice (*Oryza sativa* L.), wheat (*Triticum aestivum* L.), maize (*Zea mays*

L.), and cassava (*Manihot esculenta* L.). Sweet potato also ranks 7th most important staple food crops in the world.

Asia is the world's largest sweet potato-producing region, with 114 million tons of annual production. Japan produces about 1,300,000 ton of sweet potato annually, ranking 5th in sweet potato production in the world (Kazutoshi, 2004). Haiti annually produces 761, 000 tons of roots and tubers; of which 577, 000 tons are used for human consumption and 184, 000 tons for feed and planting materials (FAO, 2006b). Based on the importance of the root crops produced in Haiti, the ranking, in a descending order, is cassava, sweet potato, yams (*Dioscorea* spp.), potato (*Tuberosum esculenta* L.), taro (*Colocasia esculenta* Krantz), and coco yam (*Xanthosoma sagitifolia* L.).

2.1.3. Use and nutritional value

According to CGIAR (2006), sweet potato is used as staple food for human consumption, as livestock feed, and in industrial processes for making alcohol, starch, and products such as noodles, candy, desserts, and flour. Nearly half of the sweet potatoes produced in Asia are used for animal feed, with the remainder primarily used for human consumption, either as fresh or processed products. Sweet potato is rich in carbohydrates and vitamin A and can produce more edible energy per hectare per day than wheat, rice or cassava. The yellow-orange flesh cultivars are also rich in vitamins A and C. In Africa, most of sweet potato is cultivated for human consumption, and it is now being used to combat a widespread vitamin A deficiency that results in blindness or even an estimated death of 250,000-500,000 children a year.

Recently, breeding efforts have been made to improve the contents of the functional components such as anthocyanins, β-carotene, polyphenols, dietary fibers, vitamins and minerals in sweet potato. New cultivars with higher anthocyanin content that were released have been demonstrated to possess antioxidation functions. As leaves and stems of sweet potato are rich in vitamins, minerals and polyphenols, cultivars suitable for eating fresh as a green vegetable or as processed food have been also developed. However, in spite of its high nutritional value, sweet potato crop has been associated with poverty and considered as "food for the poor", according to a report jointly published by Kenyan Ministry of Agriculture (KMA), German Technical Cooperation (GTZ) and FAO (1998).

2.1.4. Sweet potato cultivars

Hundreds of sweet potato genotypes are being used throughout the world; some are unique to countries or to even smaller regions within countries. About the year 2000, the CIP sweet potato gene-bank maintained 5,526 cultivated *I. batatas* accessions from 57 countries (Zhang *et al.*, 2000). Of which, genotypes 2589 were from Latin America, specifically Peru alone contributed 1099 accessions (Huaman and Zhang, 1997). Being nearly half of the genotypes in the CIP genebank from Latin America supports the hypothesis that Central America is the primary center of diversity and most likely the center of origin of sweet potato. In industrialized countries where sweet potato is grown, only a limited number of cultivars is known to exist.

In Taiwan, sweet potato is an important crop for food and feedstuff, according to the Council of Agriculture (COA, 2006). From 1989 to 2006, the total area under the cultivation of sweet potato in Taiwan was maintained at 100 acres mainly for human consumption and food processing. Now in Taiwan, it is not only used for cooking, but also developed as a all-purpose raw material for processing into industrial products or food or feedstuff. Breeding sweet potato has historically been a focal point in agricultural research in Taiwan. In 2006, COA reported that more than 50 new cultivars had been subject to breeding during 1956-2003. The 10 most important cultivars of sweet potato grown in Taiwan up to the year 2006 were TNG57, Tainong 64, Tainong 65, TNG66, Tainong 67, Tainong 68, Tainong 69, Tainong 70, Tainong 71, and Tinong 72.

The major cultivars of sweet potato bred in Taiwan for cooking are TNG57, TNG66, Tainong 68, Tainong 70, and Taoyuan No.1. TNG73, a red flesh sweet potato evaluated in this study, was denominated in 2007. Tainong 71 and Taoyuan No. 2 have been bred for their high quality leaves for consumption. In addition,

Tainong 44 and Tainan No. 14 have been bred for amylum and Tainong 65 has been mentioned by COA for its high protein content, up to 6%. In contrast to Taiwan, many cultivars of sweet potato, though known to be grown in Haiti, are not officially registered at the Haiti Ministry of Agriculture, making their nomenclature difficult at the national level. Thus, an effort to characterize and register all the sweet potato cultivars in Haiti is urgently needed.

2.1.5. Sweet potato and food and nutrition security

A brief review of some concepts used in the food security system and analysis of the system in its global context are presented in this section. Attempts were also made to explore the actual and potential roles of sweet potato in health promotion, disease prevention, and food security. The term "food" is referred to as any substance, usually comprised primarily of carbohydrates, lipids, water and/or proteins, that can be eaten or drunk by animals (including humans) to grow, sustain health or for pleasure. English-speaking countries usually define four categories of substances as food: a) any substance, intended to be, or reasonably expected to be, ingested by humans; b) water and other drinks; c) chewing gum; d) substances used as ingredients in the preparation of food (United Kingdom Office of Public Sector Information, 1990; US Food and Drug Administration (FDA, 2006); European Parliament, 28 January 2002).

Food security is a flexible concept as reflected in the many attempts at defining it in research and policy usage (FAO, 2003). When there are not enough foods in a population for everyone to eat sufficiently to meet the needs of their body, such population is said food insecure (FAO, 2004). Food insecure is often connected with famine resulting from starvation, which involves the absence of food in entire communities. Rationing is sometimes used to distribute food in times of shortage, most notably during times of war (McGee, 2004). Starvation is a significant international problem. Besides starvation, insufficient food causes nearly a third of all babies born worldwide to die prematurely or have disabilities (Howe and Devereux, 2004).

Food deprivation is regarded as a deficit in Maslow's hierarchy of needs and is measured using famine scales (WFP, 2006). Actually, more than three billion people are affected by hunger; of which, 0.9 billion suffer from deficiency of calories and protein, 126 million children are underweighted because of inadequate intake of food and frequent disease, and more than 2 billion people suffer micronutrient deficiency due to deficiency of vitamins and minerals in the diet (FAO, 2006b). Food security is still globally a very big issue (Benson, 2004; Rosegrant *et al.*, 2005; Benson, 2006). It is estimated that 14% of children are born with low birth-weights every year (Alderman *et al.*, 2004). The poverty estimates also increased in percentage terms, with 41.6% below the poverty line in 1981, 44.6% in 1990, and 46.9% in 2001 (World Bank, 2004).

Actually, over 146 million children all over the developing countries are underweighted, representing more than one-quarter (27%) of all children under fives in the developing world (UNICEF, 2006).

Actually, food surpluses and hunger are co-existed or alternating each other. How will the situation be in the following 10 to 20 years? According to International Food Policy Research Institute (IFPRI, 1995), the trend of the world population increased from 1 billion people in 1820 to 2 billion in 1930, 3 billion in 1960, 4 billion in 1974, 5 billion in 1988, and 6 billion in 2000. It is expected to reach 7.5 billion in 2020 and nearly 9 billion in 2050. This trend of the population increase creates an undue pressure on food supply in order to meet an ever-increasing demand for food in the world.

IFPRI (1995) projected the global demand, between 1993 and 2020, for cereals to be increased by 41% or 2,490 million metric tons, for meat to be increased by 63% or 306 million tons, and for roots and tubers to be increased by 40% or 855 million tons. Currently, more than 1.3 billion people are absolutely poor, with incomes of one dollar a day or less per person, while another 2 billion people are only marginally better off (World Bank, 1997a). Due to food not being equally distributed to all, a large proportion of the population is likely to have access to less food than needed. So the problem of hunger is likely to persist.

Pinstrup-Andersen *et al.* (1997) foresaw that there will still be 167 million malnourished children in the world in 2010 and 150 million in 2020.

To eat properly being the first need of all human, it is an imperative to investigate all possible ways to impede hunger to persist in the world for long term (2020 Africa Conference Advisory Committee, 2004). Nowadays, the rich-country research agendas are shifting. In particular, they are no longer as interested in simple productivity enhancement; dietary patterns and other priorities change as incomes increase (Pardey *et al.*, 2006). Therefore, the most effective assistance that developed countries can provide to developing countries is to create good jobs with good perspective for the future.

In view of the problems of food deficiency and proverty of farmers in developing countries, attention should be paid to producing easy-to-grow and high yielding crops such as sweet potato, cassava, sorghum, and bread fruit.

Producing more than 133 million tons annually on a harvested area of about 91000 km^2 (CIP, 1998), sweet potato is an important crop to combat hunger in the developing world. It has higher yielding ability and higher energy content than potato, yam, rice, wheat, and cassava (FAO, 1984). The crop has also edible protein content comparable to yam and rice and higher than cassava. In terms of production value/ha/day, it ranks the third after potato and yam and before rice, groundnut in shell, lentil, and wheat. Its high yielding ability makes it a potential food crop to cut hunger in the world.

Having good adaptability to diverse environmental conditions and able to grow with no synthetic pesticides and fertilizer, it is suitable for economically and environmentally friendly crop, and organic and sustainable food production. According to Heidhues *et al.* (2004), adequate access to food needed for food security must be complemented with provision of health services, education, sanitary environments, and safe water sources, among other resources, to achieve nutrition security. Eating sweet potato can promote the health status and prevent some diseases. Some cultivars of sweet potato have anti-carcinogenic properties and/or AOA (Teow *et al.*, 2007). Islam *et al.* (2002) reported that its extracts effectively prevented HIV virus from proliferation. Of course, not all the cultivars are equally rich in antioxidants, but it is always possible to do a screening and select the richest cultivars for nutrition security purposes.

2.2. Food constituents and analysis

2.2.1. Introduction

Among the extremes optimal health and death from starvation or malnutrition, there is an array of disease status that can be caused or alleviated by changes in diet (Staessen *et al.*, 1999; Kessler, 2000; Kazantzis, 2004; Raisz, 2005; Prince, 2006). Deficiencies, excesses, and imbalances in diet can produce negative impacts on health, which may lead to diseases such as obesity and osteoporosis, as well as psychological and behavioral problems. Nutrients in food are grouped into several categories. Macronutrients mean carbohydrates, proteins, and lipids. Water is also a very important constituent of foods. Minor or micronutrients or secondary metabolites include mineral components and organic materials such as vitamins, enzymes, and emulsifiers. The food constituents occur naturally, their content and arrangement give foods their structure, texture, flavor, color, nutritive value, and keeping qualities (Potter, 1986). In some instance, foods also contain substances that can be toxic if consumed in large amounts. Sometimes, not satisfied with the structure, texture, flavor, color, nutritive value, or keeping qualities of a food, we add other materials that can be natural or synthetic.

2.2.2. Macronutrients

2.2.2.1. Carbohydrates in foods

Carbohydrates or saccharides are molecules that are straight-chain aldehydes (-CH_2OH) or ketones (-CO-) with many hydroxyl (OH) groups added, usually one on each carbon atom that is not part of the aldehyde or ketone functional group. The most important are sugars, starches, celluloses, hemicelluloses, pectins, and certain gums. Carbohydrates make more than 90% of the total dry matter of plants (BeMiller and Whistler, 1996). Chemically, they usually contain only carbon, hydrogen, and oxygen. Dietitians and nutritionists commonly classify carbohydrates as simple (monosaccharides and disaccharides) or complex (oligosaccharides and polysaccharides).

Carbohydrates provide 4 kcal/g, but they are not essential nutrients in terms of energy provider because the body can obtain all its energy from protein and lipids (Westman, 2002). FAO and WHO (2002) jointly recommended that national dietary guidelines set a goal of 55-75% of total energy from carbohydrates, but only 10% should be from free sugars, i.e. mono and disaccharides added to foods by the manufacturer, cook or consumer, in addition to sugars naturally present in honey, syrups, and fruit juices. Indigestible carbohydrate-components of foods like cellulose, hemicellulose, pectins, lignin, and other plant substances not readily digested are collectively referred to as fiber or dietary fiber. They hold water, tend to soften stools, and decrease stool transit time trough the large intestine. They thereby contribute to healthy conditions of the intestine.

2.2.2.2. Proteins in foods

Proteins are large molecules built from small units known as amino acids. In biology and biochemistry, amino acids are seen as the building blocks of proteins, and 20 amino acids make up all the proteins (Damodaran, 1996).

Proteins are used for growth and repair, also for strengthening the bones. They are essentials to all life, and 1 g prvides 4 kcal. National and international organizations, which advise on nutrient requirements, suggest standards that are calculated to meet or exceed the requirements of practically everyone in the population (Lappé, 1976; FAO/WHO/United Nation University (UNU), 1985; Milward *et al.*, 1992).

2.2.2.3. Lipids in foods

Lipids are a large group of fatty acid esters generally soluble in organic solvents, but only sparingly soluble in water (Nawar, 1996). There are three forms of lipids: phospholipids, steroids, and triglycerides. They are used as basic building blocks of biological membranes and for energy storage, as 1 g of lipids provides 9 kcal. Lipids consist of a polar (hydrophilic) head and one to three non-polar (hydrophobic) tails. They are called amphipathic. The hydrophobic tail consists of one or two fatty acids. Fatty acids are chains of carbon atoms connected by single bonds alone (saturated fatty acids) or by both single and double bonds (unsaturated). The chains are usually 14-24 carbon groups long. For lipids present in biological membranes, the hydrophilic head is from one of three different groups: glycolipids, phospholipids, or sterols. Glycolipids have heads containing an oligosaccharide with 1-15 sugar residues. Phospholipids possess heads containing a positively charged group that is linked to the tail by a negatively charged phosphate group. Sterols heads contain a planar steroid ring, such as cholesterol (only in animals).

2.2.2.4. Water in food

Water is a major component of some foods (Fennema, 1996). It is present in most natural foods at the extent of 70% or greater of their weight. Fruits and vegetables may contain 90% or even 95% water (Potter, 1986). Cooked foods from which some water has been removed still contain about 60% water. Water greatly affects the preserving quality of foods, which is the reason for removing it. As liquid in foods, water is very active, acting as solvent for numerous food chemicals, and thus promoting chemical reactions between dissolved constituents. Water is also necessary for microbial growth. Another reason for removing water from foods, besides preservation, is to reduce the weight and the bulk of the food; thus reducing packaging and shipping costs. However, Potter (1986) reported that, in some case, the storage ability of a dehydrated food is enhanced by leaving a trace of moisture equivalent to a monomolecular layer of water coating on all internal and external surfaces. This monomolecular layer of water serves as barrier between oxygen and sensitive constituents in the food that, otherwise, would be more easily oxidized.

2.2.3. Micronutrients

Other groups of substances play important role, out of their relatively small concentrations in foods (Potter, 1986). These constituents include natural emulsifiers, organic acids, oxidants and antioxidants, enzymes, pigments and colors, flavors, natural toxicants, vitamins, and minerals.

2.2.3.1. Natural emulsifiers

Generally, an emulsifier is any ingredient used to bind together normally non-combinative substances, such as oil and water (Potter, 1986). Emulsifiers keep fat globules dispersed in water or water droplets dispersed in fat. Without emulsifiers, mayonnaise would be separated into water and oil layers. Egg yolk contains a natural emulsifier (lecithin) and is used to thicken and bind sauces (such as hollandaise), as well as to bind ingredients in baking. Xanthan gum is a commercial emulsifier used in numerous foods like salad dressings and dairy products.

2.2.3.2. Organic acids

Fruits contain natural acids like citric acid of oranges and lemons, malic acid of apple, and tartaric acid of grapes (Potter, 1986). These acids give the fruits tartness and slowdown bacterial spoilage. Food manufacturers deliberately ferment some foods with desirable bacteria to produce acids and give the food flavor and keeping quality. Examples are fermentation of cabbage to produce lactic acid, and yield sauerkraut and fermentation of apple juice to produce first alcohol and lactic acid to obtain vinegar. In the manufacture of cheese, a bacterial starter is added to milk to produce lactic acid. This aids in curd formation and subsequent preservation of the curd against undesirable bacteria.

2.2.3.3. Oxidants and antioxidants

Oxygen that is always present in and around foods is an oxidant that can make fat, oil, and oily flavor become rancid when they are excessively exposed to air (Potter, 1986). Oxygen accelerates oxidation of these materials. Certain metals like copper and iron catalyst the oxidation process. Many natural foods contain traces of copper and iron and oxidize easily. To minimize the oxidation process, food manufacturers use nitrogen or vacuum packaging. Food processors replace copper and iron equipments by stainless steel.

Foods also contain antioxidants that tend to prevent oxidation. Natural antioxidants in foods include lecithins that are also emulsifiers; vitamins A, E, and C; and phenolics. More than 4,000 compounds in foods act as antioxidants (Kendall, 2000). They include vitamins (e.g. vitamins A, E, C) and vitamin co-factors; minerals, (e.g. selenium); hormones (e.g. melatonin); carotenoids (e.g. α-carotene and β-carotene); phenolics (e.g. anthocyanins, phenolic acids and their esters like gallic acid, salicylic acid, curcumin, etc.); and other organic antioxidants (e.g. citric acid, oxalic acid). Antioxidants are discussed in details in section 2.2.6.

2.2.3.4. Enzymes

Enzymes are biological catalysts that promote a wide variety of biochemical reactions. They are large protein molecules that, like other catalysts, need to be present in small amounts to be effective (Whitaker, 1996). Amylase found in saliva promotes digestion or breakdown of starch in the mouth. Pepsin found in gastric juice promotes digestion of proteins. Lipase found in liver promotes breakdown of fats. Thousands of enzymes exist in bacteria, yeast, molds, plants, and animals.

2.2.3.5. Pigments and colors

Pigments are natural substances in cells and tissues of plants and animals that impart color (Von Elbe and Schwartz, 1996). For example, chlorophylls impart green color to lettuce and peas, carotene gives the orange color to carrots and corn, lycopene contributes to the red color of tomato and watermelon, anthocyanins contribute to the red color of grape and blueberries, and oxymyoglobin gives the red color to meat. The food color can be altered during heating and drying according to a non-enzymic browning reaction that may be pyrolysis, caramelization, or Maillard reaction.

Pyrolysis involves the total loss of water from the sugar molecule and the breaking of carbon-carbon linkages, i.e the destruction of the sugar molecule. The result is a burnt with inedible flavor. Caramelization is a heat induced transformation of reducing sugars alone in a concentrated solution called "anhydro sugars". In this reaction the simpler sugars lose water molecules from their structure. This process is influenced by pH. The Maillard reaction is a chemical reaction between an amino acid and a reducing sugar, usually requiring the addition of heat. It is a form of non-enzymatic browning. This reaction is the basis of the flavoring industry, since the type of amino acid determines the resulting flavor. The Maillard reaction is responsible for many colors and flavors in foodstuffs like toasted bread, malted barley as in malt whiskey or beer, self-tanning products, roasted or seared meat, dried or condensed milk, etc.

The food color is measured by the mean of a colorimeter, e.g., CIELab colorimeter (CIELab, 2003). The machine reads the test sample and gives the results expressed as color parameters, e.g. L, a, and b in CIELab colorimeter. L = degree of lightness, a = red color when its value is positive (+) and green color when its value is negative (–), b = yellow color when it is positive value (+) and blue color when it is negative (–). Some derivate statistics can be computed and allow a better appreciation of the color. The angle arctangent (b/a) expressed as degrees and called hue angle (h°) as well as the chroma value (C) are usually calculated. A hue angle close to 0° or 360° expresses a red color. Near 90°, it expresses a yellow color and close to 270°, it means a blue color. Chroma value $= \sqrt{a^2 + b^2}$ expresses the measurement of hue intensity.

2.2.3.6. Flavors

Flavor is the sensory impression of a food or other substance, and is determined mainly by the chemical senses of taste and smell. Some scientists suggest it also includes sight, feeling, and sound (Lindsay, 1996). Flavorant is defined as a substance that gives another substance flavor, altering the characteristics of the solute, causing it to become sweet, sour, tangy, etc. Smell is the main determinant of a food item's flavor. While the taste of food is limited to sweet, sour, bitter, salty, umami, and other basic tastes, the smells of a food are potentially limitless.

2.2.3.7. Vitamins and minerals

Vitamins are organic chemicals other than essential amino acids and fatty acids, which a given living organism requires in trace quantities for good health, but which the organism cannot synthesize, and therefore must obtain from its diet (Gregory III, 1996). More than two billion people (i.e. one in three persons worldwide) suffer from micronutrient deficiency (hidden hunger). Vitamin D is the only major vitamin that the human body can manufacture. Vitamins are divided into two groups: fat soluble vitamins (vitamins A, D, E. K) and water soluble vitamins (vitamin C and several members of vitamin B complex).

From the 90 naturally occurring elements in the earth's crust, 25 are known to be essential to life and are present in living cells (Miller, 1996). Calcium (Ca), phosphorus (P), magnesium (Mg), sodium (Na), potassium (K), sulfur (S), and chloride (Cl) are considered as major or macro minerals in foods. Micro-minerals or trace elements include at least iron (Fe), iodine (I), zinc (Zn), selenium (Se), chromium (Cr), copper (Cu), fluorine (F), lead (Pb), tin (Sn), nickel (Ni), cobalt (Co), molybdenum (Mo), vanadium (V), arsenic (As), silicon (Si), and manganese (Mn).

2.2.3.8. Natural toxicants

Toxicants refer to inherent toxicants and contaminants in foods (Pariza, 1996). Certain mushrooms have poisonous properties due to specific nitrogen-containing bases or alkaloids that, depending on the concentration, can produce marked physiological effects. Many other natural foods also contain substances that can be harmful if consumed in large quantity. I personally already saw some mammals (pigs, goats, cows) die after consuming some high cyanide content cassava cultivars. Soil and water normally contain potentially harmful metals, as lead, mercury, cadmium, arsenic, zinc, and selenium. Traces of these metals occur in all foods, but at their low concentration of occurrence, not only are these natural components of food harmless, but zinc, selenium, and possibly others are essential to our life.

In addition of the heavy metals, some of the better known toxicants occurring naturally in foods include low levels of alkaloid such as solanine in potato and cyanide-generating compounds in lima bean (*Phaseolus lunatus* L.) and cassava.

2.2.4. Proximate analysis of foods

Proximate analysis of foods is to analyze the major food components: moisture, crude proteins, total lipids, total ash, total fiber, and total carbohydrates. We usually also include in proximate analysis tables energy as total calories. Food moisture is usually determined by gravimetric method. Moisture in a weighed sample is removed by heating up to constant weight in oven. The weight loss is calculated and reported as moisture (Meloan and Pomeranz, 1980). Many methods of protein determination in foods exist. Lowry's method (Lowry *et al.*, 1951) using bovine serum albumen (BSA) as standard for low protein products as reported by Cooper (1977) and Kjeldahl's method for both low and high protein products are very common. Kjeldahl's method has been described in details by many authors with some slight variations.

Lipids determination is done by the direct method of the American Oil Chemist Society-Association of the Official Analytical Chemists (AOAC, 1970). Around 5 g sample (w_s) is extracted with 150 ml diethyl ether for 4 hours in a thimble (soxthlet extraction). The extract is collected in a previously dried up to constant weight (w_1) round bottle and evaporated. The round bottle with the extract is then dried at 105°C up to constant weight (w_2). Drying to constant weight takes 3 to 4 hours. Crude fat is calculated as follows: $\%crude\ fat=100*(w_2-w_1)/w_s$. Food ash content is determined by weighting the dry mineral residue of organic materials heated at elevated temperature, around 550°C (Meloan and Pomeranz, 1980). An aluminum dish is dried to constant weight (A) at 100°C. The weighting is done after cooling in dessicator for 10 min. Approximately, 2 g of sample (B) is weighted in the above aluminum dish and dried at 550°C to constant weight (C). It takes approximately 6 hours. The % ash is calculated using the following formula: %ash=100*(C-A)]/B.

Crude fiber is analyzed according to AOAC methods (AOAC, 1970). In these methods, crude fiber is defined as the loss on incineration of dried residue remaining after digestion of sample with 1.25% H_2SO_4 and 1.25% NaOH solutions under specific conditions. Carbohydrate (CHO) determination is usually done by difference according to Shimelis *et al.* (2006). Total CHO (%) = 100 % – [% water + % protein + % lipids+ % ash]. Available CHO (%) = % total carbohydrates- % fiber. Energy in foods is expressed as kilocalories (kcal). 1 g of carbohydrate or protein provides 4 kcal, while 1 g of fat provides 9 kcal (Potter, 1986). Total kcal in 100 g food = 1 kcal * (% CHO*4 + % protein*4 + % fat*9).

2.2.5. Concepts of functional foods and secondary metabolites in foods

Functional foods are defined as any food or food ingredient that may provide a health benefit besides the traditional nutrients it contains. Those products, which are known as functional foods, may contain either bioactive constituents such as antioxidant vitamins in a higher quantity than present in corresponding conventional foods or may have a reduced level of undesirable components such as saturated and trans fatty acids (National Council for Agricultural Research of the Netherlands, NRLO (2001)). Biologically active components in functional foods may impart health benefits or desirable physiological effects (Goodman *et al.*, 2004). Carotenoids, precursors of vitamins A, constitute a good example. Given their structure ($C_{40}H_{54}(OH)_2$ for xanthophylls and $C_{40}H_{56}$ for carotenes), carotenoids are efficient free-radical scavengers and they enhance the vertebrate immune system (Burton and Ingold, 1984; Burton, 1989; Krinsky 1989; Palozza and Krinsky, 1992). Carotenoids are also found to be capable of inhibiting growth of certain cancer cells such as the colon cancer (van Poppel, 1993; Goh, 1996; Steinmetz and Potter, 1996).

It is common knowledge that vitamin A deficiency is a major public health problem in developing countries. Preschool children and women of reproductive age are predominantly affected (WHO/UNICEF, 1995). For these population groups, the primary source of vitamin A is derived from pro-vitamin A carotenoids in plant foods. Food-based strategies for combating vitamin A deficiency include promotion of a higher production and consumption of dark-green leafy vegetables, yellow and red fruit and vegetables, and red palm oil, which are all rich in pro-vitamin A carotenoids (Bloem *et al*, 1998).

2.2.6. Antioxidants in foods: concept, role, and assessment

2.2.6.1. Concepts and importance

The body metabolism produces usually some molecules with one or more unpaired electrons called free radicals. Those free radicals react rapidly with other molecules, starting chain of reactions in a process called oxidation. The oxidation reactions have some damaging effects in tissues and provoke some pathological conditions (Slater, 1989; Jenner, 1991; Riemersma *et al.,* 1991; Sies, 1991; Aruoma, 1993; Yoshikawa, 1993). Antioxidants in food refer to components that can counteract the damaging effects of free radicals in tissues of the consumer. Plant derived antioxidants such as vitamin E, vitamin C and phenolics have been increasingly suggested as important dietary antioxidant factors that may help to protect against oxidative damages (Mathews-Roth, 1985; Tonuci *et al.*, 1995).

Antioxidants are also well known food preservatives (Li *et al.*, 1999). Some artificial antioxidants such as butylated hydroxyanisole (BHA), butylated hydroxytoluene (BHT), *n*-propyl gallate (PG), monoglyceride citrate (MGC), and *tert*-butylhydroquinone (TBHQ) have been developed and widely used in food preservation. It was commonly admitted in the scientific area that if these chemicals are used properly and in approved quantity, they can not be harmful. However, nowadays, because of some toxic and carcinogenic concerns about artificial antioxidants and chemical food preservatives, consumers have clearly and increasingly expressed their preference for natural food additives and great interest has been placed in research on natural antioxidants (Loliger, 1991; Block *et al.*, 1992; Shahidi and Wanasundara, 1992; Stavric *et al.,* 1992; Brownridge, 1993; Hertog *et al.,* 1993; Aruoma, 1994; Bonorden and Pariza, 1994; Laranjinha *et al.*, 1994).

Preservation of foods by drying is one of the earliest and simplest techniques that have been used for centuries. The principle of dehydration is that bacteria are unable to survive in the absence of moisture (Kuntz, 1992). Sun drying is the most common method of dehydration. Besides sunlight, other methods such as cabinet dryer and oven have been used. Papad, macaroni, dried milk powder, dehydrated soups, pre-cooked legumes, and cereals are some of the commonly used dried preserved food items. Dried foods can be transported easily due to its lightweight and volume. However, during drying, vitamin C and β-carotene, two important antioxidants, are lost mainly while sugar and fiber concentrate (Rodriguez-Amaya, 2001). This method also causes severe loss of color, taste, aroma, and texture.

2.2.6.2. Antioxidants and health promotion

The most studied antioxidants are vitamins C and E, β-carotene and the mineral selenium (Kendall, 2000). Vitamin C is a water-soluble vitamin found in all body fluids. It is a powerful antioxidant that cannot be stored by the body. Vitamin E, a fat-soluble vitamin that can be stored with fat in the liver and other tissues, is promoted for a range of purposes from delaying aging to healing sunburn. Another antioxidant, β-carotene, which is most extensively studied among over 600 different carotenoids, has been reported to protect dark green, yellow, and orange vegetables and fruits from solar radiation damage. β-carotene is thought to play a similar role in protecting animal organisms from the solar radiation damage. Selenium is a micro-mineral that is thought to help fight cell damage by oxygen-derived compounds and thus may help protect against cancer. It is best to obtain selenium through foods, as large doses of the supplement form can be toxic. In traditional societies, nutrition and health care are interconnected and many plants are consumed as food in order to benefit health (Etkin, 1996; Pieroni, 2000; Pieroni *et al.*, 2002a, 2002b; Ansari *et al.*, 2005; Pieroni and Price, 2005).

2.2.6.3. Concept of total antioxidant activity of foods

Some natural food components such as carotenoids (Higashi-Okai *et al.*, 2001), flavonoids, simple phenols, and glucobrasicins (Awad *et al.*, 2001), have become objects of interest due to their beneficial properties for health. Practically, all the above mentioned antioxidant compounds are obtained through the ingestion of plant products such as fruits and vegetables, nuts, flours, vegetable oils, drinks, and infusions. Such ingestion can be in the form of fresh produce or processed products in cans, juices, frozen, cooked, sauces, etc. Methods of analysis have been developed to verify the presence of antioxidants and their endogenous quantities in every foodstuff.

The most widely used separation, identification and quantification techniques are high performance liquid chromatography (HPLC) with photometric, electrochemical, or fluorescence detection and mass spectrometry with capillary gas chromatography (GC-MS), or liquid chromatography (LC-MS).

However, knowledge of the potential antioxidant compounds of a food does not necessarily indicate its antioxidant capacity because the cooperative effect which could exist between different antioxidants (termed synergism) can make that the total effect greater than the sum of the individual activities and the isolation of one compound would not exactly reflect the overall action (Arnao *et al.*, 1999). Additionally, unidentified and unmeasured components in the samples could potentially produce a similar effect. Thus, in food evaluation, it is difficult and not useful to measure each antioxidant component separately. For this reason, methods to quantify a parameter that expresses the total capacity of a sample (in either a fresh raw, or processed product) to scavenge free radicals in liquid media have been sought for many years (Halliwell and Gutteridge, 1990). This parameter is the total AOA that such methods look at as a global characteristic of a product, independent of its particular composition.

2.2.6.4. Strategies to measure total antioxidant activity of foods

Arnao *et al.* (1999) presented two strategies to measure total AOA. These strategies are based on the inhibition of certain reactions by the presence of antioxidants. The methods involve the generation of a free radical compound, which disappears in presence of antioxidants. Radicals are used as a quantifying or a detectable factor in the presence of antioxidant compounds to provoke a loss or an increase of signal (generally absorbance or fluorescence). Using these techniques, it is possible to define two strategies for studying the total AOA of a compound, or biological sample.

***Strategy* 1:** Discoloration or signal loss assay. The reaction produces a quantity of free radicals from the reagent(s), which are sufficiently stable that, after the addition of antioxidant, the signal loss (loss of radical) can be used as an indicator of total AOA. Quantification is done by means of a calibration curve that relates the loss of signal to the concentration of antioxidant present.

***Strategy** 2:* An inhibition assay in which the reaction course is estimated in the absence (control) or presence of different quantities of antioxidant (A1, A2...). The reagents and the antioxidant sample are mixed at the onset of the reaction. The presence of antioxidant affects the reaction in two ways: (A) Inhibition of the signal per unit of time (t). Thus, we can make a calibration curve which correlates the % of inhibition for every concentration of antioxidant used, with respect to an assay control without antioxidant. (B) By registering the course of the reaction, the appearance of a slow or *lag* phase can be observed. The duration of the *lag* phase (time before reaching the steady state) is proportional to the concentration of antioxidant in the sample.

2.2.6.5. Some selected techniques of assessment total AOA in foods

Arnao *et al.* (1999) presented some selected methods that have been developed to study the total AOA of fresh plant materials and foodstuffs. The assays using the radicals DPPH (Blois, 1958) and ABTS (Miller *et al.*, 1993; Miller and Rice-Evans, 1997; Bartosz *et al.*, 1998; Cano *et al.*, 1998; Cano *et al.*, 2003;) are among the most common of these methods. We can also mention: 6-hydroxy-2,5,7,8-tetramethylchroman-2-carboxylic acid (trolox)-equivalent antioxidant capacity known as TEAC (Wang *et al.*, 2004)**;** oxygen radical absorbing capacity, known as ORAC (Cao *et al.*, 1993; Cao *et al.*, 1995; Wayner *et al.*, 1985); and antioxidative potency in linoleic acid model system assays (AOP).

2.2.6.5.1. DPPH method

The principle of use of the stable free radical DPPH for estimating AOA was introduced by Blois (1958) and reviewed by many authors including Hogg *et al.* (1961) and Molyneux (2004). The molecule of DPPH is characterized as a stable free radical by virtue of the delocalization of the spare electron over the molecule as a whole, so that the molecules do not dimerise, as would be the case with most other free radicals. The delocalization also gives rise to the deep violet color, characterized by an absorption band in ethanol solution centered at about 520 nm. When a solution of DPPH is mixed with that of a substance that can donate a hydrogen atom, then this gives rise to the reduced form with the loss of this violet color (although there would be expected to be a residual pale yellow color from the picryl group still present). The DPPH free radical scavenging activity (FRSA) is computed, knowing the absorbance values of the test sample and a blank as follows: DPPH FRSA (%) $=100*(A_0-A_s)/A_0$, where A_0 is initial absorbance value for the DPPH solution before adding the test sample and As, absorbance value for the mixture DPPH with the test sample (Molyneux, 2004). Higher antioxidant contents correspond to lower absorbance values and higher DPPH inhibition.

The working wavelength to be used for the absorbance measurements varied with different authors. For example, 515 nm has been recommended by Brand-Williams *et al.* (1995), Bondet *et al.* (1997), Sanchez-Moreno *et al.* (1999), Lebeau *et al.* (2000), and Gomez- Alonso *et al.* (2003), while Schwarz *et al.* (2001) proposed 516 nm. Blois (1958) who originally developed the DPPH method and Zhu *et al.* (2002) recommended 517 nm, whereas Leitão *et al.* (2002) proposed 518 nm, and Kim *et al.* (2002), 520 nm. Alternately, the concentration of antioxidants compounds may be determined using a standard. Suitable standards that are widely used are ascorbic acid (Brand-Williams *et al.*, 1995; Sanchez-Moreno *et al.*, 1998); α-tocopherol (Lu and Foo, 2000; Guo *et al.*, 2001); and trolox (Wang *et al.*, 2004).

2.2.6.5.2. ABTS methods

The ABTS method was developed first by Miller and Rice-Evans (1997). Then, many authors adapted it to an extent that nowadays, we can say ABTS methods (plural) instead of ABTS method (singular). In effect, ABTS is a stable free radical used in combination with different other chemicals alternately in AOA assessment. The principle remains the same, but many protocols have been developed. Bartosz *et al.* (1998) used 2,2'-azobis(2-amidopropane) hydrochloride (ABAP) as a source of peroxyl and alkoxyl radicals which oxidizes ABTS to a green cation radical. The method took then the name of ABTS/ABAP method. When antioxidants are present in a sample, the reaction is inhibited. The induction time of the reaction is used as a parameter to determine the antioxidant content.

Cano *et al.* (1998), Arnao *et al.* (1999) and Cano *et al.* (2003) used horse radish peroxidase (HRP) instead of ABAP, and they called the method ABTS/HRP. Their assays were based on the principle of inhibition of free radical generated by hydrogen peroxide from ABTS in presence of HRP. The added antioxidant provokes a reduction of the ABTS radical and a decrease of absorbance. Lana (2005 and 2006) modified the ABTS/HRP method to evaluate both hydrophilic and lipophilic AOA in tomato. Teow *et al.* (2007) used ABTS salt in combination with potassium persulfate. Bouayed *at al.* (2007) prepared the reaction medium mixing 2,2-azobis(2methylpropionamidine) dichloride (AAPH), a radical initiator, with ABTS in phosphate-buffered saline (PBS, pH 7.4). They used vitamin C as standard and they called the method: Vitamin C equivalent antioxidant capacity method (VCEAC).

2.2.6.5.3. TEAC method

The TEAC assay is also an adaptation of Miller and Rice-Evans (1997) ABTS method. It is based on the suppression of the absorbance of ABTS radical by antioxidants in the test sample when ABTS (610 μM) incubates with a peroxidase (metmyoglobin, 100 μl, 6.1 μM) and H_2O_2 (20 μl, 100 μM). A trolox standard calibration curve is drawn and used to compute the AOA as trolox-equivalent antioxidant capacity (TEAC). The trolox dose–response relationship is linear when incubation is extended beyond 24 min (Wang *et al.*, 2004).

2.2.6.5.4. ORAC method

The ORAC method is based on the decrease in fluorescence of ß-phycoerythrin (ß-PE) in presence of ABAP (Cao *et al.*, 1993; Cao *et al.*, 1995). A lag phase or rate constant for ß-PE fluorescence decay is used to determine the AOA of the added sample. Thus, it implies the use of a spectrofluorometer rather than a spectrophotometer. A trolox standard calibration curve (0~100 μM/l) is used to evaluate the AOA. Wang *et al.* (2004) found no significant difference between estimates of change in total AOA of rat plasma as measured by TEAC and ORAC methods, but the correlation between the two assays was poor.

2.2.7. Place of the solvent in antioxidant activity analyses

The analysis of total AOA in liquid media implies previous extraction of the antioxidant components by using appropriate solvents. Though supercritical fluid extraction uses gaseous supercritical fluids, a solvent is commonly regarded as an inert liquid that dissolves a solid, liquid, or gaseous solute, resulting in a solution (Houghton and Raman, 1998). The most common solvent in everyday life is water. Most other commonly-used solvents are carbon-containing chemicals called organic solvents. The use of inorganic solvents is typically limited to research in chemistry and some technological processes. Solvents usually have a low boiling point, so that they evaporate easily and can be removed by distillation, thereby leaving the dissolved substance behind. Besides their low boiling point, their polarity and their density are two important characteristics of solvents. The polarity of a solvent determines what type of compounds it can dissolve and with what other solvents or liquid compounds it is miscible.

A solvent dissolves compounds with similar polarity, and the nature of the solvent used determines the types of chemicals likely extracted from the plant material (Rodney, 2004). The density difference allows liquid-liquid partition of compounds. The result obtained in the analysis of a plant material for total AOA depends on the solvent used for extracting the antioxidant compounds.

2.2.8. Phenolic compounds in foods

Polyphenolic compounds or phenolics are ubiquitous secondary metabolites in plants. They comprise of a large group, above 8000, biologically active substances from simple phenol molecule to polymeric structures with molecular mass above 30000 Da (Marinova *et al.*, 2005). On the basis of the number of simple phenol units, they are classified into two basic groups, simple phenols and poly-phenols. The group of simple phenols contains also the so-called phenolic acids or phenols with carboxylic groups and that of poly-phenols contains at least two phenol rings in their structures. Flavonoids are the major subgroup of polyphenols. Phenolic acids can be classified in benzoic acids and their derivatives and cinnamic acids and their derivatives.

The main subclasses of flavonoids are flavonols (catechins and related compounds; and quercetins and their related glycosides), flavones (apigenin and luteolin), flavanonas (naringin) and isoflavones (genistein and daidzein). All of them mostly occur in nature as glycoside structures (Blasco *et al.*, 2005). More than 5000 flavonoid structures are currently known in higher and lower plants, indicating the chemical complexity of these compounds. According to Kurata *et al.* (2006), sweet potato leaves have a high content of phenolics responsible for its anti-carcinogenic properties.

2.2.8.1. Total phenolic content assessment

The presence of phenolic compounds in diet was considered in the past as a negative feature because they were doubted to make the nutrients less available (Marinova *et al.*, 2005). Nowadays, particular emphasis is put on studying phenolic compounds because data showed that the presence of phenolics in food is important for their oxidative stability and antimicrobial protection (Awad *et al.*, 2001; Ehlenfeldt. and Prior, 2001). Total phenolics in foods are assessed usually by Folin-Ciocateu assay. This assay consists of a reaction of phenolic compounds with Folin-Ciocalteu phenol reagent in presence of Na_2CO_3. Folin-Ciocalteu phenol reagent does not contain phenols, but it reacts with phenols in presence of Na_2CO_3 by developing color. The color intensity depends on the phenol content in the reaction medium. A gallic acid standard calibration curve is usually prepared by plotting gallic acid solution concentrations (10 to 400 mg/l) against their absorbencies at wavelength about 750~765 nm. Total phenolic content of food extracts are computed substituting the absorbencies of mixture of such food extracts with Folin-Ciocalteu phenol reagent in presence of Na_2CO_3 in the standard calibration equation.

2.2.8.2. Total flavonoids content assessment

The total flavonoids content is assessed by the aluminum trichloride colorimetric method. Many adaptations of this method have been done. Marinova *et al.* (2005) prepared standard samples mixing an aliquot (1 ml) catechin standard solutions (20, 40, 60, 80, 100 mg/l) with 4 ml ddH_2O, 0.3 ml 5% $NaNO_2$, 0.3 ml 10% $AlCl_3$, and 1 ml 1 M NaOH. They completed to volume (10 ml) with ddH_2O. They read absorbencies at 510 nm after mixing. They prepared the test sample replacing 1 ml catechin solution by 1 ml plant extract.

Miliauskaus *et al.* (2004) assessed the content of flavonoids using rutin as standard. A rutin standard solution was prepared by diluting 0.05 g rutin in ethanol and completing to volume (25 ml) with ethanol. An aliquot (1 ml) plant extract (10 g/l) in methanol was mixed with 1 ml (20 g/l) $AlCl_3$ in ethanol and diluted with ethanol to 25 ml. The absorption at 415 nm was read after 40 min at 20°C. The amount of flavonoids in plant extracts in rutin equivalents (RE) was calculated by the following formula: X=[10 (A / Ao) x (Mo / M)] where *X*= flavonoid content as mg RE /g dry plant extract; *A*= absorbance of plant extract solution; *Ao*= absorbance of rutin standard solution; *M*= weight (g) of plant extract in the sample giving absorbance A, and *M*o=0.05 g, weight of rutin in the sample giving the absorbance Ao. Huang *et al.* (2004) mixed 1.5 ml extracts with equal volumes of a solution of methanolic (2%) $AlCl_3\text{-}6H_2O$ solution. The mixture was vigorously shaken, and absorbance at 367 nm was read after 10 min incubation. Flavonoid content was expressed as mg catechin equivalent (CE)/ 100 g fresh matter.

2.2.9. Carotenoids in foods

2.2.9.1. Occurrence and role

Carotenoids belong to the group of yellow and red pigments that occur widely in plants and some other photosynthetic organisms like algae, some types of fungus, and some bacteria. They interact with ROS and act in animal organisms as free radical scavengers and lipid antioxidants (Palozza and Krinsky 1992). The number of known natural carotenoids was approximately 80 in 1960 (Zechmeister, 1960), 500 in 1980 and more than 600 at the beginning of 2000s (Kendall, 2000). According to Ladilav *et al.* (2005), since 1980, around 7500 papers had already been published about carotenoids. Food carotenoids are C_{40} molecules built from eight isoprenes (C_5H_8) units (Rodriguez-Amaya, 2001). Carotenoids with molecules containing oxygen, such as lutein, are known as xanthophylls (originally phylloxanthins). Their chemical structure is $C_{40}H_{54}(OH)_2$. The oxygen-free carotenoids such as α-carotene and β-carotene, are known as carotenes. Carotenes typically contain only carbon and hydrogen, with the formula $C_{40}H_{56}$. The common food carotenoids can be either acyclic like phytoene, phytofluene, and ζ-carotene or cyclic like ß-zeacarotene, α-zeacarotene, and γ-carotene.

Alpha-carotene is more widespread, but usually present at low levels except in Brazilian passion fruit (Mercadante *et al.* 1998) and carambola (Gross *et al.*, 1983) where it occurs as a major pigment. Lycopene is the main pigment of many red-fleshed fruits and fruit vegetables such as tomato, watermelon, red-fleshed papaya and guava, and red or pink grapefruit. ß-carotene is the most widespread carotenoid in foods, either as a minor or major pigment (e.g., apricot, carrot, mango, West Indian Cherry). α-carotene and the γ-carotene sometimes accompany ß-carotene, but at lower concentrations (Rodriguez-Amaya, 2001).

Substantial amounts of γ-carotene are found in carrot and some cultivars of squash and pumpkin (Arima and Rodriguez-Amaya, 1988, 1990) and substantial amounts of δ-carotene are found in rose hips, *Eugenia uniflora,* high delta strain of tomato, and the peach palm fruit. Less frequently encountered is δ-carotene, although it is the main carotenoid of the high delta strain of tomato and the peach palm fruit. Crude palm oil is the richest source of carotenoids in nature in terms of retinol equivalent with 500-700 ppm (Yap, 1995; Puah *et al.*, 2005). It contains 11 types of carotenoids, of which α-carotene and ß-carotene account for more than 91%. Carotenoids are found in carrots and responsible for their bright orange color (Hendren *et al.*, 2002).

According to Rodriguez-Amaya (2001), apricot, carrot, mango, loquat, West Indian cherry, and palm fruits are rich in ß-carotene. The xanthophylls α-cryptoxanthin and zeinoxanthin are reported by Rodriguez-Amaya (2001) to occur widely in plants, although generally at low levels. Leaves have a strikingly constant carotenoids pattern, often referred to as the chloroplast carotenoids pattern, the main carotenoids being lutein (about 45%), ß-carotene (usually 25–30%), violaxanthin (15%), and neoxanthin (15%) (Britton, 1991). α-Carotene, ß-cryptoxanthin, α-cryptoxanthin, zeaxanthin, antheraxanthin, and lutein 5,6-epoxide are also reported as minor carotenoids. Some palm fruits (e.g., *buriti*) are especially rich in carotenoids, particularly provitamin A carotenes. According to Bovell-Benjamin (2007), a large range of natural products including carotenoids exist in sweet potato. Carotenoids have many physiological functions. They are efficient free-radical scavengers (Burton and Ingold, 1984; Burton, 1989; Krinsky 1989) and they enhance the vertebrate immune system.

The ability of carotenoids to quench singlet oxygen is related to the conjugated double-bond system, and maximum protection is given by those having nine or more double bonds (Foote *et al.*, 1970). The acyclic lycopene was observed to be more effective than the bicyclic ß-carotene (Di Mascio *et al.*, 1989). Results obtained with a free radical–initiated system also indicated that canthaxanthin and astaxanthin, in both of which the conjugated double-bond system is extended with carbonyl groups, were better antioxidants than ß-carotene and zeaxanthin (Terão, 1989). Carotenoids are also found to be capable of inhibiting growth of certain cancer cells such as the colon cancer (Goh, 1996).

Structurally, vitamin A or retinol ($C_{20}H_{30}O$) is essentially one half of the molecule of ß-carotene with an added molecule of water at the end of the lateral polyene chain.

It is common knowledge that vitamin A deficiency is a major public health problem in developing countries. Preschool children and women of reproductive age are predominantly affected (WHO/UNICEF, 1995). For these population groups, the primary source of vitamin A is derived from provitamin A carotenoids in plant foods. Food-based strategies for combating vitamin A deficiency include promotion of a higher production and consumption of dark-green leafy vegetables, yellow and red fruit and vegetables and red palm oil, which are all rich in provitamin A carotenoids (Bloem *et al*, 1998). Some epidemiological studies have suggested a lowered risk of developing certain types of cancer (Goh, 1996) or chronic diseases following a diet containing food products rich in carotenoids, due to their antioxidative properties (Block *et al*, 1992; van Poppel, 1993; Steinmetz and Potter, 1996).

2.2.9.2. Carotenoids analysis

Carotenoids analysis includes extraction, partition, saponification, and washing, concentration or evaporation of solvent, chromatographic separation, identification, and quantification. Protection from light, avoiding high temperature, avoiding contact with acid, and use of high purity solvents that are free from harmful impurities are recommended (Davies, 1976; Britton, 1991; Schiedt and Liaaen-Jensen, 1995). Extraction of carotenoids is usually done with water-miscible organic solvent, such as acetone, methanol, ethanol, or mixtures thereof to allow better solvent penetration (Rodriguez-Amaya, 2001). Prolonged soaking should, however, be avoided to prevent isomerization and degradation of the carotenoids. Most carotenoids are stable towards alkali.

A neutralizing agent (e.g., calcium carbonate, magnesium carbonate, or sodium bicarbonate) may be added during extraction to neutralize acids liberated from the food sample itself (Rodriguez-Amaya, 2001). Fractions or isolates should be kept dry under nitrogen or argon or dissolved in a hydrocarbon solvent, petroleum ether or hexane, and kept at –20°C or lower when not in use (Craft and Soares, 1992). Partition is done by liquid-liquid extraction, also known as solvent extraction and partitioning. The extract of fresh materials usually contains a substantial amount of water, which can be removed by partition to hexane, petroleum ether, diethyl ether, or dichloromethane or mixtures of these solvents.

In the case of carotenoids, diethyl ether or a mixture of ether with hexane or petroleum ether is preferred for extracts with large amounts of xanthophylls, part of which is lost during partition with pure hexane or petroleum ether.

Saponification aims to selectively remove the chlorophylls and esterified fatty acids from the organic phase using a strongly alkaline agent (Larsen and Christensen, 2005). However, saponification extends the analysis time, and may provoke artifact formation and degradation of carotenoids. Therefore, it should be omitted from the analytic procedure whenever possible. When indispensable, saponification is best done after transferring the carotenoids to petroleum ether or hexane, by adding an equal volume of 10% methanolic potassium hydroxide with the addition of butylated hydroxytoluene (Rodriguez-Amaya, 2001). The mixture is left overnight at room temperature in darkness, after which the carotenoid solution is washed 5 times with water to remove the alkali. Concern about the possible negative effects of saponification has recently led researchers to shorten the duration of ambient temperature saponification, for example 1 or 2 h.

In HPLC methods, the extract is often evaporated to dryness and then dissolved in the mobile phase or a solvent compatible with the mobile phase. Carotenoids extract or solution should be concentrated in a rotary evaporator at reduced pressure and a temperature ideally below 40ºC. Care should be taken to prevent the extract from going to complete dryness in the rotary evaporator because this may result in degradation of carotenoids, especially lycopene (Tonucci *et al.,* 1995). Additionally, part of the carotenoids (especially the more polar ones), may adhere strongly to the glass walls, precluding quantitative removal from the flask. Separation of carotenoids can be done by open column chromatography (OCC) or HPLC.

However, Rodriguez-Amaya (2001) reported that reproducibility of the results is one of the main problems in chromatographic separation. The variations tend to be greater in developing countries where quality control of laboratory materials may not be rigorous. Therefore, a laboratory's first attempt may not duplicate reported separation, and adjustment of the chromatographic conditions may be necessary. Conclusive identification is a requirement for the accurate determination of the carotenoid composition of foods. Ideally, the identification procedure should include mass spectrometry (MS), especially electron-impact MS, which gives the molecular ion and fragments characteristic of structural features of the carotenoid molecule (Rodriguez-Amaya, 2001).

Electron impact–MS was used recently to confirm the identity of carotenoids from mango and yellow passion fruit (Mercadante *et al.*, 1997; Mercadante *et al.*, 1998). Quantification of carotenoids is done using standards. Unfortunately, only a few carotenoid standards (e.g., α-carotene, β-carotene, and lycopene) are available commercially. The accuracy of the analytic results depends on how accurately the concentrations of the standard solutions are known. However, commercial β-carotene standards have been shown to have widely varied purity (Quackenbush and Smallidge, 1986; Craft *et al.,* 1990). Both OCC and HPLC methods assume that tabulated absorption coefficients are accurate, which is not the case for all carotenoids included in the tables. Inaccuracies in the preparation of the solutions, determination of the concentrations, and construction of the calibration curves will affect the reliability of the results.

Khachik *et al.* (1992) cited the following parameters for evaluating the validity of the standards and the instrumentation: the correlation coefficient should be greater than 0.9, the intercept should be very close to zero, and the relative standard deviation of the regression (standard error of the estimate divided by average concentration of standards multiplied by 100) should be less than 5%. If any of these parameters is out of range, the standard as well as the HPLC instrumentation should be carefully examined and the standard curve should be rerun.

2.3. Conclusions

Sweet potato production and utilization exceeded 133 million tons of tuberous roots annually in the world with 95% of which produced in developing countries. With more than 5000 genotypes in the world, it indicates that sweet potato is a plant extensively used for food, feed or other products. It provides more energy per ha and per day than cassava and wheat. Some genotypes have been reported to possess AOA and anticarcinogenic properties. Those findings are encouraging as more research efforts will be made to evaluate greater number of sweet potato genotypes and develop innoving techologies for processing and storage of sweet potato. Education of consumers on its nutrional and functional values can free sweet potato from the stigma of being a "food for poors" and contribute to the enhancement of its production, utilization and consumption. This will ultimately improve the health of the poor populations in the world and ensure the global food and nutrition security. Study of the AOA of sweet potato is part of this effort.

PART III. EXPERIMENTATION

The experiments, carried out to achieve the overall objective identified above, were divided into seven major parts and each was reported separately in seven chapers (chapters 3 to 9). The first experiments conducted were to evaluate three cultivars of sweet potato (*Ipomoea batatas* Lam.); TNG57, TNG66 and TNG73; by comparing their color characteristics, proximate composition, and carotenoid contents. In the second part of the experimentation, the antioxidant activity (AOA) and the total phenolic content of the extracts from raw sweet potato roots were analyzed. The effects of the extracting solvent (solvent optimization) and the genotype on the levels of the AOA and the total phenolic content as well as the distribution pattern of the AOA and the total phenolic content between sweet potato flesh and skin were also studied. The results of the solvent optimization test were used to choose the most appropriate solvent for further phenolic antioxidant extraction. In the third part, the changes in the AOA and the phenolic content of the three cultivars during the freeze-drying process were evaluated using non-dried samples as control.

Similarly, the experiments conducted in the fourth part studied the effects of the four air drying treatments; namely sun-drying, low-temperature air drying (LTD) at 25°C, hot air drying (HAD) at 50ºC, and HAD at 75°C; on the AOA and the total phenolic, flavonoid, carotenoid, and beta-carotene contents of TNG57, TNG66, and TNG73, using freeze-drying as control. To analyze the AOA in these experiments, three analytical methods were employed, including 1, 1-diphenyl-1-picrylhydrazyl (DPPH), 2,2'-azino-di-[3-ethylbenzthiazoline-6-sulfonate] (ABTS), and anti-oxidative potency in linoleic acid system model (AOP) assays. Comparisons were made to determine the efficiency of these three methods in measuring the AOA. The experiments in the fith part were conducted to identify the major phenolic antioxidants of TNG73.

In the sixth part, flesh and skin samples of TNG73 were processed into flours and stored for 12 months at room temperature (25ºC) and 4ºC; experiments were carried out to evaluate, during the 12 month-storage, the changes in the AOA and total phenolic content of the flours. The experiments in the seventh part analyzed the pasting profile of flours and starches isolated from the tuberous roots of TNG57, TNG66, and TNG73.

Figure 1 shows a diagram of the relationship of the response variables and their related factors tested in each of the seven set of experiments conducted in this study.

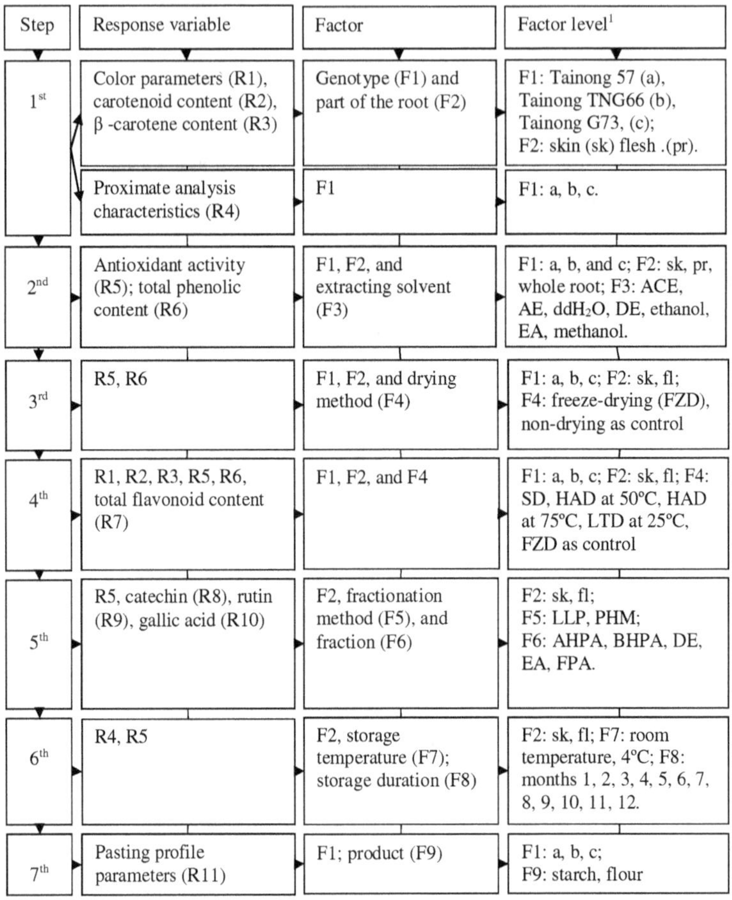

Figure 1. Diagram of the study design.

[1]ACE: acetone, AE: aqueous ethanol (1:1, v/v), ddH_2O: distilled de-ionized water, DE: diethyl ether, EA: ethyl acetate; SD: sun-drying, HAD: hot air drying, LTD: low temperature air drying; LLP: liquid-liquid partition, PHM: pH modification; AHPA: acid hydrolysable phenolics, BHPA: base hydrolysable phenolics, FPA: free phenolics.

Chapter 3. Color characteristics, proximate composition, and carotenoid content of TNG57, TNG66, and TNG73 sweet potato (*Ipomoea batatas* Lam.) cultivars

Abstract

Color characteritics, proximate composition, and carotenoid content in the tuberous roots of the three cultivars of sweet potato (*Ipomoea batatas* Lam.); TNG57, TNG66, and TNG73; were evaluated. The color was measured by the colorimeter using the *L, a, b* scale. The total carotenoid content was determined spectrophotometrically, and the beta-carotene content was analyzed by HPLC. Of the three cultivars, TNG57 was yellow, while TNG73 skin and flesh were yellowish red and red, respectively. TNG66 skin was yellowish red, while its flesh was reddish yellow. The crude fiber content was ranging from 1.88 to 2.19% on the dry matter basis, while the energy content was between 383.51 and 385.51 kcal/ 100 g of dry matter. TNG66 exhibited the highest contents of crude ash, proteins, lipids, and carotenoids, but with the lowest content of available carbohydrates. TNG57 and TNG73 had similar contents of crude ash, lipids, and carbohydrates. Among the three cultivars, TNG73 showed the lowest content of crude proteins and carotenoids, which, in an ascending order, was followed by TNG57 and TNG66.

Key words: carotenoid content, color parameters, nutritious composition, sweet potato

3.1. Introduction

Color parameters are important variables in AOA analysis because the AOA activity of foodstuffs are usually associated with some particular pigments like phenolics, carotenoids, betalains, or clorophylls, etc (Elbe and Schwartz, 1996). The color parameters of the fresh samples of three cultivars of sweet potato, TNG57, TNG66, and TNG73, were analyzed to serve as a baseline reference for determining the drying effects on the color and the AOA. The proximate analysis is a minimum analysis required for food items. The objective of the proximate analysis in this study was to provide with basic information on the major nutritious components of the of TNG57, TNG66, and TNG73. Their moisture, crude ash, total fiber, crude protein, total lipids, total carbohydrate, and total calorie contents were analyzed. This information is essential to meeting the label requirements on any processed and packaged food. In addition to the general composition, the total carotenoid and beta-carotene contents of the samples were also analyzed.

Carotenoids have many physiological functions. They interact with ROS and act as efficient free-radical scavengers and lipid antioxidants in animal organisms (Palozza and Krinsky 1992). They enhance vertebrate immune system (Krinsky, 1989) and are capable of inhibiting growth of certain cancer cells such as the colon cancer (Goh, 1996). The number of known natural carotenoids was approximately 80 in 1960 (Zechmeister, 1960), 500 in 1980 and more than 600 at the beginning of 2000s (Kendall, 2000). According to Ladilav *et al.* (2005), since 1980, around 7500 papers have already been published about carotenoids. Some cultivars of sweet potato have been reported to possess high carotenoid content (Teow *et al.*, 2007). ß-carotene, which is aboundant in fruits such as apricot, carrot, mango, and West Indian Cherry, is the most widespread carotenoid in foods, either as a minor or major pigment (Rodriguez-Amaya, 2001). Structurally, vitamin A or retinol ($C_{20}H_{30}O$) is essentially one half of the molecule of ß-carotene with an added molecule of water at the end of the lateral polyene chain.

It is a common knowledge that vitamin A deficiency is a major public health problem in developing countries, with preschool children and women of reproductive age being predominantly affected (WHO/UNICEF, 1995). For these population groups, the primary source of vitamin A is derived from provitamin A carotenoids in plant foods. Food-based strategies for combating vitamin A deficiency include promotion of a higher production and consumption of high provitamin A carotenoid crops (Bloem *et al.*, 1998). Therefore, TNG57, TNG66, and TNG73 were evaluated and compared for their total carotenoid and beta-carotene contents.

3.2. Materials and Methods

3.2.1. Materials

Freshly harvested sweet potato tuberous roots were purchased from a local market in Pingtung, Taiwan, in March 2007. Ammonium sulfate, beta-carotene, boric acid, color indicator (methyl red and methyl blue mixture 2:1), concentrated H_2SO_4 (95-98%), diethyl ether, hydrogen chloride (HCl), potassium sodium tartrate, sodium acetate x $3H_2O$, sodium carbonate, standard beta-carotene, sodium hydroxide (NaOH), and catalyst were purchased from Sigma Chemical Co. (St. Louis, MO, USA). All the other reagents were either extra pure or analytical grade.

3.2.2. Experiment design

Completely randomized design was used with the three genotypes as treatments. The proximate analysis was conducted in ducplicate for the flesh samples, while the color parameters and carotenoid content were measured in triplicate for both the skin and the flesh samples. Figure 2 is an illustration of the experiment design for proximate analysis.

3.2.3. Color parameter measurements

The sample color parameters were evaluated in a Color Quest II – Sphere Colorimeter ("Hunter Lab Reston", VA) using the L, a, b scale (CIELab, 2003). The L value is the degree of lightness of the sample, varying from 0=black to 100=white. The a value means red color when it is positive (+) and green color when it is negative (–). The b value corresponds to yellow color when it is positive (+) and blue color when it is negative (–) (Higby, 1962). The hue angle = arctangent (b/a), and chroma (C) $=\sqrt{a^2+b^2}$ were calculated using Excel software. Excel returns the arctangent (b/a) in radian (h^{rd}), and the results were transformed into degree (h^o) using the following formula: $h^o = 180 \times h^{rd}/\pi$, where π=3.1416. The chroma value is a measurement of the hue intensity.

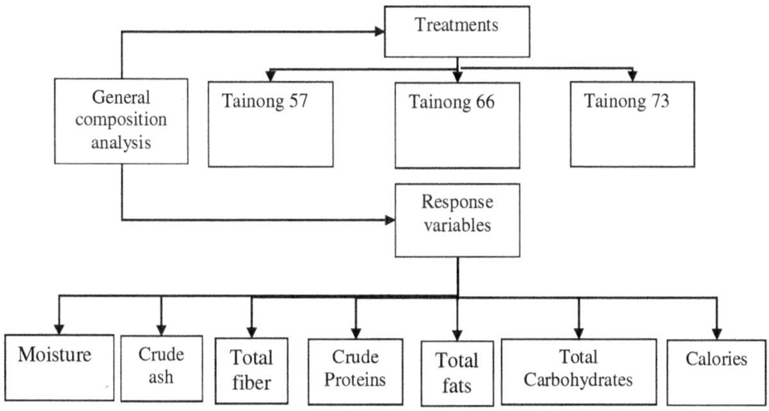

Figure 2. Illustration of the experiment design for proximate analysis conducted in step one of the research project.

3.2.4. Proximate analysis

The proximate analysis was performed by measuring moisture, crude ash, total fiber, crude proteins, total lipids, available carbohydrates, and calories. Moisture content of the samples was determined according to a gravimetric indirect method of the Association of Official Analytical Chemists (AOAC, 1970) as described by Meloan and Pomeranz (1980). An aluminum dish was dried to constant weight (A) at 100°C. The weighing was done after cooling in desiccators for 10 min. An aliquot of 5±0.1g of fresh sample (B) was weighted in the above aluminum dish and dried at 100°C to constant weight (C). The sample dry matter (SDM) was calculated using the following formula: SDM (%)=[(C-A)/B]*100. The moisture content (%) was obtained by subtracting the SDM from 100%.

The ash content was determined by weighing the dry mineral residue of the organic materials heated at elevated temperature, around 550°C, applying the method 936.07of AOAC (2000). An ashing dish was dried up to constant weight (A) at 100°C. An aliquot of 2±0.1 g of sample (B) was weighted in the above ashing dish and ignited in furnace at 550°C for 6 h (to constant weight, C). The total ash content was calculated using the following formula: % ash=[(C-A)]/B]*100.

Crude fiber was analyzed by using the method 962.09 of AOAC (2000). An aliquot of 2 g dried sample (A) was digested consecutively in 200 ml boiling 1.25% H_2SO_4 and 200 ml boiling 1.25% NaOH. The remaining indigestible residue (fiber+ash) was dried at 110°C to constant weight (B), cooled in desiccators, and incinerated in electric muffle furnace at 550°C. The weight after incineration (C) was measured and recorded. The crude fiber was defined as the weight loss on incineration of the dried residue remaining after digestion of the sample. Crude fiber (%)=[(B-C)/A]*100.

Crude protein analysis was performed by using Kjeldhal block digestion and steam distillation, applying the official 990.02 prescribed AOAC (2000). This method consists of three steps: digestion of the nitrogen in a digestion unit including digestion tubes and heating block digester, distillation in a distillation unit including distillation flasks, and titration. A reverse-reading burette was used for dispensing standard NaOH, HCl and H_2SO_4. The content of nitrogen obtained was converted into crude protein content using a conversion factor of 6.25.

The direct method 920.85 of AOAC (2000) was used to analyze the total lipid content. Around 5 g sample (w_s) was extracted with 150 ml diethyl ether for 4 hours in a thimble. The extract was collected in a previously dried to constant weight (w_1) round bottle and evaporated. The round bottle with the extract residue was then dried at 105°C to constant weight (w_2). The total fat was calculated as follows: Total fat (%)=[(w_2-w_1)/w_s]*100.

Total carbohydrates (CHO) content was computed by difference according to Shimelis *et al.* (2006). Total CHO (% dry matter) = 100 % – [% protein + % lipids+ % ash]. Available CHO (%) = % total carbohydrates- % fiber. Energy in foods is expressed as kilocalories. Carbohydrates or proteins provide 4 kcal/g, while lipids provide 9 kcal/g (Potter, 1986). Therefore, energy (kcal/100 g) = (total cabohydrates*4 + total proteins*4 + total lipids*9).

3.2.5. Carotenoids extraction

The extraction of fresh sweet potato for total carotenoids and beta-carotene analysis was conducted according to Teow *et al.* (2007). Briefly, after cleaning, peeling, and cutting, a 25 g sample of sweet potato was blended with 2 g Na_2CO_3, 1 g diatomaceous earth, 25 ml methanol, and 50 ml hexane-acetone (1:1 v/v). The mixture was stirred for 2 h in the dark, and filtered through Whatman no. 1 filter paper (Whatman Inc., Clifton, NJ). The carotenoids dissolved in the solvent mixture passed through the filter paper. The solid residue was collected and washed twice with 25 ml methanol and once with 50 ml hexane-acetone mixture. The filtrates were combined in a separatory funnel, diluted to 250 ml with ddH_2O, and kept at room temperature in the dark until phase separation. After phase separation, the aqueous phase (lower layer) was discarded, the upper layer was transferred into a 50 ml volumetric flask, made to 50 ml with hexane, and used for analysis.

3.2.6. Total carotenoids and beta-carotene analysis

The absorbance of the carotenoid extracts was read in triplicate in spectrophotometer at 450 nm (Teow *et al.*, 2007). The total carotenoid content was calculated according to the following formula (Gomes, 2007; Higby, 1962):

$$CT = \frac{A_{450}\ x1000}{250\ xLxW},$$

where CT is the total carotenoid content in mg/kg fresh matter; A_{450}, the absorbance of the extract at 450 nm; L, the path-length of the cuvette used in the spectrophotometer (1 cm); w, the initial amount of sample (25 g fresh sample) divided by the final volume of extract obtained (50 ml). The results were converted into mg/kg dry matter.

Identification and quantification of beta-carotene were conducted by HPLC according to Ribeiro *et al.* (2007). An aliquot of 2 ml extract was filtered through a 0.45 μm Millipore membrane, and a 20 μL sample was injected in duplicate into the liquid chromatographic system consisting of a high-pressure pump (HITACHI L-7100, Japan) and a UV-visible detector (HITACHI L-7420, Japan). Separation was carried out in a RP18 column (Microsorb, 5 μm, 4 nm × 250 nm). Wavelength detection was 450 nm and the mobile phase consisted of methanol, EA, and acetonitrile (70:20:10) at a flow rate of 1.0 ml/min. Quantification was done using standard *beta*-carotene calibration curve. The HITACHI Chromatography Data Acquisition Software was used to collect and process the data. The peak area was used to compute the beta-carotene content in mg/l. The results were converted into mg/kg fresh matter and mg/kg dry matter.

3.2.7. Statistical analysis

The data were subjected to the analysis of variance according to Kuehl (1999). The SAS (Statistical Analysis System, v. 8.1; SAS Institute Inc., Cary, NC, USA) software was used. Treatment effects were evaluated by F test at $\alpha=0.05$. The results were reported as mean±SE. Means were compared by Duncan multiple range test. A difference at $p<0.05$ was considered as significant.

3.3. Results

3.3.1. Color parameters

The hue angle values listed in Table 1 and the visual color appearances shown in Figure 3 indicated that the color of the TNG57 samples was yellow both in the skin and the flesh, but significantly lighter in the flesh. The flesh of TNG66 samples was light reddish yellow while the skin was dark yellowish red. We commonly refer to reddish yellow and yellowish red as orange. In TNG73, the flesh was red and the skin yellowish red. Unlike in TNG57 and TNG66, the color of TNG73 samples was darker in the flesh than the skin. In the three cultivars, the color was more intense in the flesh than the skin.

3.3.2. Proximate composition

Results of the proximate analysis of the samples are summarized in Table 2. The crude ash content varied from 2.36 % dry matter in TNG57 to 3.43 % dry matter in TNG66. The % crude fiber varied from 1.88% in TNG57 to 2.19% in TNG73, while it was 1.98% in TNG66. The % crude proteins significantely varied from 2.45 % dry matter in TNG73 to 8.96 % dry matter in TNG66, whereas TNG57 had an intermediate content of 4.49% dry matter. The crude lipids content in TNG73, TNG66 and TNG57 was 0.47%, 1.43% and 0.27%, respectively. The energy content was approximately the same in the three cultivars (384.39, 385.5, and 383.51 kcal /100 g dry matter in TNG57, TNG66, and TNG73, respectively). TNG66 exhibited the highest contents of crude ash, proteins, and lipids, but the lowest content of available carbohydrates. TNG57 and TNG73 had similar contents of crude ash, lipids, and carbohydrates. TNG73 showed the lowest content of crude proteins and, in an ascending order, was followed by TNG57 and TNG66.

Table 1. Hue angle, color lightness, and color intensity of the fresh samples of sweet potato cultivars TNG57, TNG66, and TNG73

Part of the root	(Mean±SE, n=3)[1]		
	Hue angle (h°)	Lightness (L)	Chroma (C)
	----------TNG57----------		
Flesh	86.51±0.56[a]	71.62±1.40[a]	27.93±2.04[a]
Skin	75.96±2.18[b]	53.62±1.86[b]	19.57±0.37[b]
	----------TNG66----------		
Flesh	56.73±2.63[a]	63.40±1.60[a]	34.33±0.93[a]
Skin	43.56±5.07[b]	25.36±2.68[b]	15.84±1.39[b]
	----------TNG73----------		
Flesh	354.54±0.88[a]	25.36±1.55[b]	34.98±0.47[a]
Skin	33.61±5.87[b]	34.07±1.25[a]	9.17±0.59[b]

[1]Means with different letters in a column within cultivars are significantly different ($p<0.05$).

Figure 3. Color appearance of entire and cut roots of TNG57, TNG66, and TNG73.

A, B, C: entire roots of TNG57, TNG66, and TNG73, respectively.

D, E, F: cut roots of TNG57, TNG66, and TNG73, respectively.

Table 2. General composition of 100 g sample dry sweet potato flesh

Component	(Mean±SE, n=2)[1]		
	TNG57	TNG66	TNG73
Crude ash (g)	2.36±0.02[b]	3.43±0.01[a]	2.52±<0.01[b]
Crude fiber (g)	1.88±0.05[a]	1.98±0.05[a]	2.19±0.05[a]
Proteins (g)	4.49±0.02[b]	8.96±0.08[a]	2.45±<0.01[c]
Crude fats (g)	0.27±0.01[b]	1.43±0.02[a]	0.47±<0.01[b]
Avail.[2] carbohydrates (g)	91.00±0.52[a]	84.20±0.29[b]	92.37±0.60[a]
Energy (kcal)	384.39±3.57[a]	385.51±3.72[a]	383.51±2.97[a]

[1]Means with different letters in a row are significantly different ($p<0.05$). [2]Avail: available.

3.3.3. Total carotenoids and beta-carotene contents

The total carotenoids and beta-carotene contents of the three cultivars of sweet potato were analyzed by spectrophotometry and HPLC, respectively. Figure 4 shows representative chromatograms of standard beta-carotene and a TNG66 extract. Table 3 shows that TNG66 had the highest content of total carotenoids with 510.39 μg beta-carotene equivalent/g dry matter (156.61 mg beta-carotene equivalent/kg fresh matter) and beta-carotene with 464.61 mg/kg dry matter (142.59 mg beta-carotene equivalent/kg dry fresh matter) than TNG57 and TNG73. The total carotenoid content of TNG57 was 133.52 mg beta-carotene equivalent/kg dry matter (39.05 mg beta-carotene equivalent/kg fresh matter), while its content of beta-carotene was 131.7 mg/kg dry matter corresponding to 35.56 μg beta-carotene equivalent/g fresh matter. TNG73 exhibited the lowest content of total carotenoids with 32.04 mg beta-carotene equivalent/kg dry matter (8.46 mg beta-carotene equivalent/kg fresh matter), while its beta-carotene content was not detected.

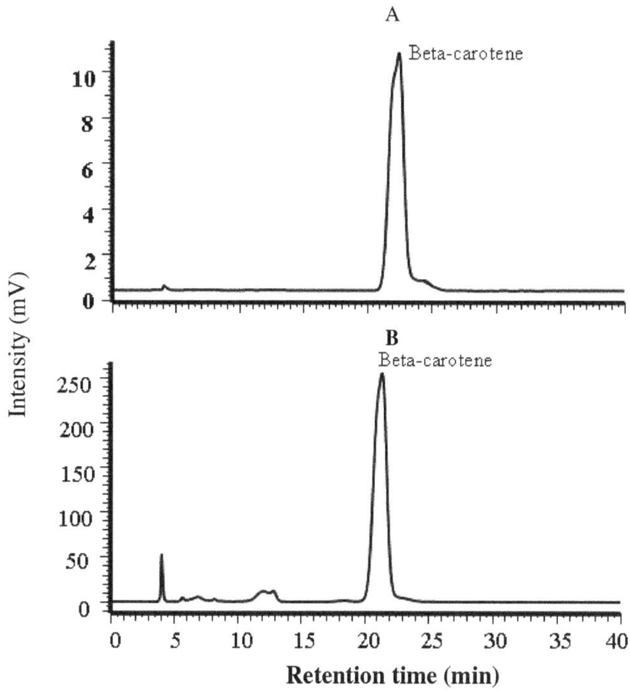

Figure 4. Representative chromatograms of beta-carotene and TNG66 extract.

A: standard, B: hexane extract of TNG66.

Table 3. Total carotenoids and beta-carotene contents of TNG57, TNG66, and TNG73

Cultivar	(Mean±SE, n=3)[1]	
	Total carotenoids (mg B-carotene equivalent/kg dry matter)	Beta-carotene (mg/kg dry matter)
TNG57	133.52±0.33[b]	131.70±0.02[b]
TNG66	510.39±0.71[a]	464.61±29.78[a]
TNG73	32.04±0.04[c]	ND

[1]Means with different letters in a column are significantly different ($p<0.05$); ND: not detected.

3.4. Discussion

TNG57, TNG66, and TNG73 show intense colors on their flesh from yellow to red. These colors suggested the results of substantial contents of phenolics or carotenoids found in the flesh of sweet potato; deep yellow to orange colors indicating high contents of carotenoids, particularly beta-carotene, and red color indicating high content of phenolics (Teow *et al.,* 2007). The carotenoid and phenolic contents of the three cultivars were further evaluated.

TNG66 exhibited the highest contents of total ash, crude proteins, and crude lipids, while the total ash and crude lipid contents of TNG57 and TNG73 were not significantly different. TNG73 showed the lowest content of proteins, and was followed in ascending order by TNG57 and TNG66. The three cultivars of sweet potato were not significantly different in terms of fiber content and energy. Huang *et al.* (2004) reported that crude protein content of sweet potato usually varies among cultivars between 1 and 3%, but rarely reaches 10% on a dry matter basis. Based on this, TNG66 may be considered particularly rich in proteins.

In studying different cultivars with different color on the flesh, Teow *et al.* (2007) observed large variation in carotenoid contents of sweet potato (0.18 to 226 μg beta-carotene equivalent/g fresh matter). Based on the results reported by Teow *et al.* (2007), TNG66 with 510.39 beta-carotene equivalent/g fresh matter may be considered as a cultivar with high carotenoid content. It was followed in descending order by TNG57 and then TNG73. Hence, THG66, with its high protein and beta-carotene contents, is suitable for being grown as a crop to meet the needs of food and nutrition security for the people particularly in areas pronged to have vitamin A deficiencies.

3.5. Conclusions

The flesh colors of the three cultivars of sweet potato used in this study were distinctively different. The orange flesh color TNG66 had the highest contents of ash, proteins, lipids, and carotenoids, but the lowest content of carbohydrates among the three cultivars studied. Because it exhibited exhibited high in content of protein and carotenoid, TNG66 is suitable for mass propagation for food and nutrition security purposes, particularly in areas pronged to have vitamin A deficiencies.

Chapter 4. Total phenolics and antioxidant activity of sweet potato (*Ipomoea batatas* Lam.) as affected by the extracting solvent and the genotype

Abstract

The effects of seven solvents were studied on the total phenolic content and antioxidant activity (AOA) of extracts of three sweet potato (*Ipomoea batatas* Lam.) cultivars: TNG57, TNG66, and TTNG73. Peeled root, unpeeled root, and skin samples were extracted with acetone, aqueous ethanol (1:1 v/v), diethyl ether (DE), distilled de-ionized water (ddH$_2$O), ethanol, ethyl acetate (EA), and methanol. The total phenolics of the extracts were assayed by Folin-Ciocalteu method and the AOA were analyzed by 1, 1-diphenyl-2-picrylhidrazyl (DPPH) assay. The ethanol extracts exhibited the highest phenolic content and AOA, and the skin of the samples had in average higher phenolic content and DPPH free radical scavenging activity (FRSA) than the flesh. TNG73 had the highest phenolic content, followed by TNG66, and then TNG57. However, the DPPH FRSA did not vary significantly among the three cultivars when extracted with ethanol 95 %.

Key words: antioxidant activity, DPPH assay, phenolics, carotenoids, sweet potato

4.1. Introduction

The body metabolism produces molecules with unpaired electrons called free radicals. Those free radicals are involved in oxidation reactions, damaging the cells. Their on the tissues are referred to as oxidative stress that leads to pathological disorders (Diaz *et al.*, 1997). The human body also produces other molecules that can counteract the damaging effects of free radicals. Such molecules are called antioxidants. The AOA of a food or a medicinal product is its ability to counteract oxidative stress. On the other hand, food and nutrition security has always been and will be, according to the projections, for longtime a big challenge (FAO, 2006a).

Besides food production and food distribution, food storage has been recognized as a key factor in addressing the issue of food and nutrition security. To delay the onset of deteriorative oxidation processes during storage, many food additives have been synthesized and widely used in food industry (Li *et al.*, 2006). However, concerns about toxicity and carcinogenicity of the synthetic food additives have been going increasing. In parallel, secondary metabolites like carotenoids, phenolics, chlorophylls, betalains, etc., known for protecting plants in their natural environments against adverse conditions have been found to increase the storage life of foodstuffs and to counteract oxidative stress in animal tissues (Bendich and Olson, 1989). Nowadays, it is commonly admitted that eating a diet rich in antioxidant-containing foods, has been linked to a reduced risk of some pathological conditions (Tonucci *et al.*, 1995; Mathews-Roth, 1991; Krinsky, 1994).

Consequently, consumers' demands for natural food ingredients have continuously increased and there has been a general desire to replace synthetic food additives with natural options (Aruoma, 1994; Hertog *et al.*, 1993). Health promotion and diseases prevention by appropriate diets have increasingly drawn attention of the scientific community. Studies have suggested that antioxidants occur naturally in many fresh foods, particularly fresh fruits, vegetables, and whole grains, and great interest has been placed in screening and ranking plant materials and food commodities for total AOA. Methods of analysis have been developed to verify the presence of a certain antioxidant compounds and their endogenous quantities in every foodstuff.

A perusal of different papers on antioxidant compounds shows that the most widely used separation, identification, and quantification techniques are high performance liquid chromatography (HPLC) with photometric, electrochemical, or fluorescence detection and mass spectrometry with capillary gas chromatography (GC-MS), or liquid chromatography (LC-MS). However, knowledge of the potential antioxidant compounds of a food does not necessarily indicate its AOA because the synergism effect which could exist between different antioxidant compounds can make the total AOA effect greater than the sum of the individual activities and the isolation of one compound would not exactly reflect the overall action (Arnao *et al.*, 1999). Additionally, unidentified and unmeasured components in the samples could potentially produce a similar effect.

Therefore, methods to quantify a parameter that expresses the total capacity of a sample to scavenge free radicals in liquid media have been sought for many years (Halliwell and Gutteridge, 1990). This parameter is the so-called total AOA that such methods look at as a global characteristic of a product, independent of its particular composition. The analysis of total AOA in liquid media implies previous extraction of the antioxidant components by using appropriate solvents.

Though supercritical fluid extraction uses gaseous supercritical fluids, a solvent is commonly regarded as an inert liquid that dissolves a solid, liquid, or gaseous solute, resulting in a solution. Some important characteristics of solvents with regard to their uses to extract antioxidant compounds are their boiling, density, and polarity. A good solvent usually has a low boiling point, so it evaporates easily and can be removed by distillation, thereby leaving the dissolved substance behind. The density is an important characteristic when an extraction involves more than one solvent and liquid-liquid partition is desirable. The polarity of a solvent determines what type of compounds it is able to dissolve and with what other solvents or liquid compounds it is miscible. A solvent dissolves compounds with similar polarity and the result obtained in the analysis of a plant material for total AOA depends on the solvent used. Antioxidant compounds vary in quality and quantity among different crops, and different extraction methods give different results (Fogliano *et al.*, 1999). Then, investigations for selecting appropriate extracting solvents are necessary in any in-depth study of AOA.

The objective of this study step was to analyze the total phenolic content and AOA of seven extracts of fresh roots of three sweet potato cultivars, namely, TNG57, TNG66, and TNG73, as influenced by the extracting solvent and the genotype. In addition, the distribution pattern of the total phenolic and AOA between the skin and the flesh was analyzed.

4.2. Materials and Methods

4.2.1. Materials

Freshly harvested sweet potato tuberous roots were purchased from a local market in Pingtung, Taiwan, in March 2007. DPPH radical, acetone, copper sulfate pentahydrate, diethyl ether, ethanol, EA, Folin-Ciocalteu reagent, gallic acid, sodium carbonate, sodium hydroxide, and vitamin C were purchased from Sigma Chemical Co. (St. Louis, MO, USA). All the other reagents were either extra pure or analytical grade.

4.2.2. Experiment design

The design consisted of 7 solvents x 3 cultivars x 3 parts of the root (peeled root, unpeeled root, and skin) factorial experiment in a completely randomized design with three replicates, yielding 63 treatments and 189 experiment units. Table 4 is the matrix of 63 treatments issued from the combination of three cultivars with three parts of the root and seven extracting solvents. These experiments aimed to optimize the extracting solvent and analyze the distribution pattern of the AOA and total phenolic content between the flesh and the skin of the three cultivars of sweet potato. Seven solvents: ethanol, methanol, diethyl ether, acetone, aqueous ethanol (1:1, v/v), ddH_2O, and EA, were used. All measurements were done in triplicate.

Table 4. Matrix of 63 treatments issued from a factorial combination of three cultivars with three parts of the root and seven solvents in the second step of the project research

Cultivar	Part of root	Solvent[1]						
		A	B	C	D	E	F	G
TNG57	Unpeeled root	01	02	03	04	05	06	07
	Peeled root	08	09	10	11	12	13	14
	Skin	15	16	17	18	19	20	21
TNG66	Unpeeled root	22	23	24	25	26	27	28
	Peeled root	29	30	31	32	33	34	35
	Skin	36	37	38	39	40	41	42
TNG73	Unpeeled root	43	44	45	46	47	48	49
	Peeled root	50	51	52	53	54	55	56
	Skin	57	58	59	60	61	62	63

[1]A: acetone, B: aqueous ethanol (1:1, V/V), C: diethyl ether, D: distilled de-ionized water, E: ethanol, F: ethyl acetate, G: methanol.

4.2.3. Crude extract preparation

An aliquot of 100±0.1 g sample (flesh, skin, or unpeeled root) was cut into small pieces (0.5 to 1 cm thickness), blended for 2 min with 400 ml solvent and transferred into a 500 ml flask. The flask was then shaken at 100 rpm in darkness for 24 h in a water bath set at 25°C for complete extraction (Auerbach and Gray, 1999; Huang *et al.*, 2004). The mixture was filtered through Whatman no. 1 filter paper (Whatman Inc., Clifton, NJ). The residue was washed with an additional 100 ml extracting solvent. The filtrate was evaporated to dryness using a rotary evaporator model BUCHI 111 equipped with a water-bath BUCHI 461 (Switzerland) and a water vacuum pump EYELA A-3S (Tokyo, Japan). The dried residue, referred to as crude extract, was re-suspended in ethanol 95%, adjusted to a concentration of 10 mg/ml, and stored at -20 °C until analysis. Figure 5 illustrates the process of fresh samples extraction.

4.2.4. Total phenolic content assay

The total phenolic assay was conducted according to the method of Folin and Ciocalteu (1927) as reported by Miliauskas *et al.* (2004) using gallic acid as standard. Briefly, an aliquot of 1 ml ethanolic gallic acid solution (0.010, 0.020, 0.050, 0.100, 0.200, or 0.400 mg/ml) was mixed in a test tube with 5 ml 2.0 M Folin–Ciocaleu phenol's reagent diluted ten-fold. After 5 min, 4 ml aqueous sodium carbonate solution (75 g/l) were added and vortex-mixed. After 60 min incubation at 20°C, the absorbance at 765 nm was read in a spectrophotometer model Beckman Coulter DU 730 UV/VIS. A standard gallic acid calibration curve ($y_{\text{(mg gallic acid/ml)}} = ax+b$, where a is the slop of the regression equation; x, the gallic acid sample absorbance at 765 nm; and b, the intercept of the equation) was drawn.

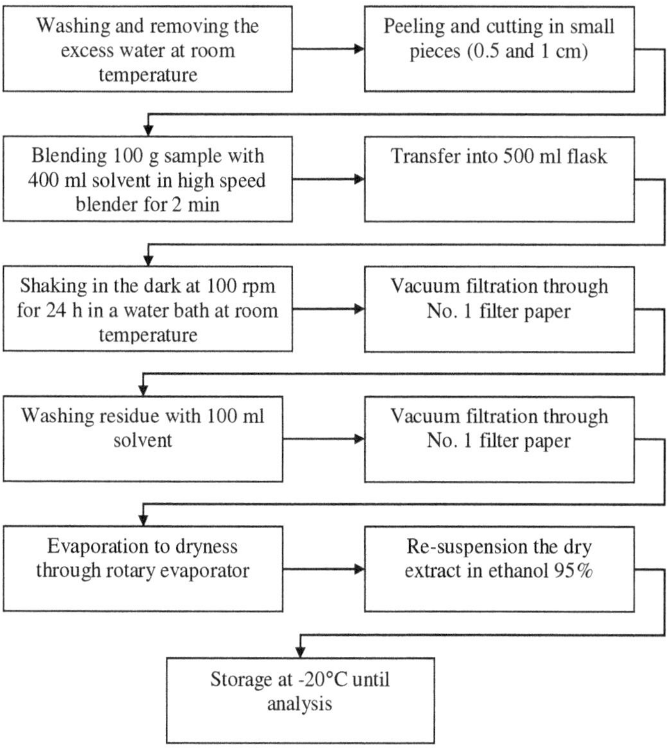

Figure 5. Flowchart of the sample extraction to measure DPPH FRSA and total phenolic content of fresh samples.

Separately, 1 ml plant extract solution (10 g/l) was mixed with 5 ml 2.0 M Folin–Ciocaleu phenol's reagent diluted ten-fold in ddH_2O. After 5 min incubation at room temperature and in darkness, 4 ml aqueous sodium carbonate solution (75 g/l) was added and vortex-mixed. After 90 min incubation at 20°C and in darkness, we read the absorbencies at 765 nm. The gallic acid standard calibration curve was used to compute the total phenolic content in mg gallic acid equivalent (GAE)/ml, replacing x by the test sample absorbance in the above equation. Then, the total phenolic content was converted into mg GAE/g dry extract.

4.2.4. Antioxidant activity assay

The total AOA was assayed by the DPPH method as described in Molyneux (2004). A 100 μM DPPH stock solution was prepared by dissolving 4 mg DPPH radical in 100 ml of pure ethanol. Then, 1.5 ml of stock solution was added to 4.5 ml of 5 mg/ml plant extract and vortex-mixed. Separately, 1.5 ml of 100 μM DPPH solution was mixed with 4.5 ml ethanol resulting in a 6 ml blank. After 30 min incubation in dark at room temperature, absorbencies of the test sample and blank were read at 517 nm in three replicates. Total AOA of the samples as DPPH FRSA was determined using the following formula: DPPH FRSA (%) $=[(A_0-A_f)/A_0]*100$, where A_0 is the blank absorbance at 517 nm after 30 min incubation and A_f, test sample absorbance at 517 nm after 30 min incubation. Additionally, different concentrations of vitamin C (0.25, 0.50, 1, 2, 4, 8, and 16 mg/l) were plotted against their DPPH FRSA under the conditions of the experiment to determine the optimum level of vitamin C that scavenged the maximum DPPH radical. This optimum level was used as reference to compare the extracts.

To analyze the effect of the incubation temperature on the DPPH FRSA, ethanol extracts were incubated at increasing (25, 50, 75 and 100 °C) for 30 min. The samples were placed in a water bath adjusted to the required temperature and tested for DPPH FRSA after 30 min (Rehman *et al.,* 2003).

4.2.6. Statistical analysis

The data were submitted to an analysis of variance according to Kuehl (1999). Table 5 shows the structure of the analysis of variance of the DPPH FRSA. The SAS (Statistical Analysis System, v. 8.1; SAS Institute Inc., Cary, NC, USA) software was used. The main factors' effects and interactions were determined by Fisher-Snedecor's F test at α=0.05. Results were reported as mean ± SE. Means comparison was done by Duncan multiple range test for main effects and by t test for interactive effects at α level= 0.05. The mathematical model used was:

$$Y_{ijkl}=\mu+\alpha_i+\beta_j+(\alpha\beta)_{ij}+\gamma_k+(\alpha\gamma)_{ik}+(\beta\gamma)_{jk}+(\alpha\beta\gamma)_{ijk}+\varepsilon_{ijkl}.$$

In this model, Y_{ijkl} represents the observed value of a dependent variable (DPPH FRSA or total phenolics); μ, the overall mean of a variable in the experiment ; α_i, the cultivar effect; β_j, part of the root effect; $(\alpha\beta)_{ij}$, the double interaction between cultivar and part of the root; γk, the double solvent effect; $(\alpha\gamma)_{ik}$, the double interaction between cultivar and solvent; $(\beta\gamma)_{jk}$, the double interaction between part of the root and solvent; $(\alpha\beta\gamma)_{ijk}$= the triple interaction between cultivar, part of the root, and solvent; ε_{ijkl}, experimental error.

Table 5. Structure of the analysis of variance (ANOVA) for DPPH FRSA of the fresh samples

Source of variation	Df[1]	SQ[2]	MS[3]	F[4]	Pr>F[5]
Cultivar					
Part of the root					
Cultivar x part of the interaction					
Solvent					
Cultivar x solvent interaction					
Part of the root x solvent interaction					
Cultivar x part of the root x solvent interaction					
Error					
Total					

[1]Degree of freedom; [2]sum of square; [3]mean square; [4]computed F-value; [5]probability for a computed F-value to be exceeded by a critical F-value.

4.3. Results

4.3.1. Total phenolic content

Among the extracting solvents, ethanol and diethyl ether showed the highest effectiveness in term of main effects on the total phenolic content of the extracts. They were followed by acetone, then methanol, then EA, then aqueous ethanol, and then ddH_2O (Table 6). TNG73 with an average of 22.73±0.95 mg GAE/g dry extract, corresponding to 114.11±4.77 mg GAE/100 g fresh matter had the highest phenolic content, followed by TNG66 (11.96±0.87 mg GAE/g dry extract, corresponding to 57.29±0.87 mg GAE/100 g fresh, and then TNG57 (11.00±1.20 mg GAE/g dry extract, corresponding to 56.65±6.18 mg GAE/100 g fresh).

The results summarized in Table 7 show that the average phenolic content of the extracts was the highest in the skin samples (19.23±1.29 mg GAE/g dry extract), followed by the unpeeled root (14.18±1.12 mg GAE/g dry extract) and then the peeled root (12.26±1.06 mg GAE/g dry extract) ones.

Significant interactions of second and third order were observed among the factors solvent, genotype, and part of the root. An example of second order interaction between solvent and genotype is the ranking of ethanol as equally, less, and more effective than diethyl ether for extracting phenolic compounds from TNG57, TNG66, and TNG73, respectively (Table 6). Solvent also interacted also with part of the root. For example, ethanol ranked the best solvent, followed by diethyl ether, for extracting peeled and unpeeled root samples, while diethyl ether ranked the best for extracting skin samples (Table 7). Another example of this second order interaction between solvent and part of the root corresponds to methanol which showed equal, higher, and lower effectiveness than acetone in extracting peeled root, unpeeled root, and skin, respectively.

Table 6. Interactive effects of the solvent and genotype on total phenolic content of the extract of TNG57, TNG66, and TNG73

Solvent[1]	(Mean±SE, n=3)[2]			
	TNG57	TNG66	TNG73	Mean
	-----mg gallic acid equivalent/g dry extract-----			
Ethanol	21.91±2.36^{a}	18.16±1.66^{b}	32.20±0.55^{a}	24.09±1.50^{a}
Diethyl ether	22.10±4.88^{a}	22.17±1.60^{a}	27.92±0.67^{b}	24.06±1.74^{a}
Acetone	7.79±1.21^{b}	11.26±2.59^{c}	25.31±0.73^{c}	14.79±1.76^{b}
Methanol	7.04±0.83^{c}	10.58±1.31^{d}	24.60±0.70^{d}	14.07±1.58^{c}
Ethyl acetate	7.68±1.77^{b}	8.19±0.13^{e}	18.84±3.41^{e}	11.57±1.59^{d}
Aq. ETOH	4.45±0.46^{e}	6.70±0.33^{f}	17.47±1.07^{f}	9.54±1.18^{e}
ddH_2O	5.99±0.28^{d}	6.56±0.30^{f}	12.74±1.01^{g}	8.43±0.69^{f}

[1]Aq. ETOH: aqueous ethanol; dH_2O: distilled de-ionized water.[2]Means with a same letter in a column are not significantly different ($p>0.05$).

Table 7. Interactive effects of the solvent, part of the root, and genotype on total phenolic content of TNG57, TNG66, and TNG73

Solvent[1]	(Mean±SE, n=3)[2]			
	TNG57	TNG66	TNG73	Mean
	-------------mg gallic acid equivalent/g dry extract-------------			
Peeled root (PR)				
Ethanol	14.45±1.70^{Ca}	17.55±0.38^{Bb}	31.47±0.42^{Aa}	21.16±2.67^{a}
DE	8.29±0.00^{Cb}	18.79±0.07^{Ba}	30.04±0.22^{Ab}	19.04±3.14^{b}
Methanol	5.51±017^{Cc}	7.37±0.02^{Bc}	22.22±0.04^{Ac}	11.70±2.64^{c}
Acetone	7.59±0.03^{Bb}	4.04±0.00^{Ce}	22.41±0.06^{Ac}	11.35±2.82^{c}
Aq. ETOH	5.26±0.03^{Bc}	5.46±0.01^{Bd}	15.67±0.15^{Ad}	8.80± 1.72^{d}
ddH$_2$O	5.25±0.16^{Cc}	7.35±0.07^{Bc}	9.41±0.26^{Ae}	7.34±0.61^{e}
EA	1.76±0.16^{Cd}	8.16±0.23^{Bc}	9.31±0.44^{Ae}	6.41±1.18^{f}
PR mean	6.87±0.84^{C}	9.82±1.22^{B}	20.08±1.87^{A}	12.26±1.06
Unpeeled root (UR)				
Ethanol	30.29±0.25^{Ba}	13.08±0.73^{Cb}	30.99±0.34^{Aa}	24.79±2.94^{a}
DE	17.08±0.11^{Cb}	19.13±0.02^{Ba}	25.47±0.11^{Ac}	20.56±1.26^{b}
Methanol	5.27±0.01^{Cd}	8.61±0.09^{Bc}	27.01±0.21^{Ab}	13.63±3.38^{c}
Acetone	3.72±0.02^{Ce}	8.46±0.09^{Bc}	26.50±0.07^{Abc}	12.89±3.47^{d}
Aq. ETOH	2.63±0.02^{Cf}	7.67±0.02^{Bcd}	15.00±0.05^{Ad}	8.43±1.80^{f}
ddH$_2$O	7.07±0.08^{Bb}	6.93±0.13^{Bd}	12.42±0.05^{Ae}	8.80±0.90^{f}
EA	7.36±0.40^{Bc}	7.87±0.20^{Bc}	15.25±0.63^{Ad}	10.16±1.29^{e}
UR mean	10.49±2.06^{B}	10.25±0.92^{B}	21.81±1.53^{A}	14.18±1.12
Skin				
Ethanol	20.98±0.45^{Cb}	23.87±1.70^{Bb}	34.13±0.65^{Aa}	26.32±2.06^{b}
DE	40.93±0.03^{Aa}	28.58±0.05^{Ba}	28.25±0.06^{Bc}	32.59±2.09^{a}
Methanol	10.33±0.027^{Ce}	15.75±0.08^{Bd}	24.58±0.11^{Ae}	16.89±2.08^{e}
Acetone	12.07±0.05^{Cd}	21.28±0.04^{Bc}	27.03±0.16^{Ad}	20.13±2.18^{c}
Aq. ETOH	5.47±0.02^{Cf}	6.98±0.04^{Bf}	21.73±0.11^{Af}	11.39±2.59^{f}
ddH$_2$O	5.66±0.02^{Cf}	5.41±0.05^{Bg}	16.39±0.09^{Ag}	9.16±1.81^{g}
EA	13.93±0.46^{Bc}	8.53±0.04^{Ce}	31.96±0.82^{Ab}	18.14±3.55^{d}
Skin mean	15.63±2.58^{B}	15.77±1.90^{B}	26.30±1.26^{A}	19.23±1.29

[1]DE: diethyl ether; Aq. ETOH: aqueous ethanol; EA: ethyl acetate; ddH$_2$O: distilled deionized water. [2]Means with a same capital letter in rows or small letter in columns within samples are not significantly different ($p>0.05$).

Second order interactions between genotype and part of the root were also observed. In effect, higher phenolic contents were observed in peeled root extracts of TNG66 than in peeled root extracts of TNG57. However, no significant difference was observed TNG57 and TNG66 in terms of phenolic contents of their unpeeled root and skin extracts. Considering TNG57, ethanol ranked as a more effective solvent than diethyl ether to extract the phenolic compounds from the peeled and unpeeled roots, while diethyl ether ranked more effective to extract the skin (Table 7). This is an example of third order interaction.

4.3.2. DPPH free radical scavenging activity (FRSA)

An optimum concentration of commercial vitamin C was determined and used as reference to compare the DPPH FRSA of the samples. That optimum concentration was found to be 16 ppm, which quenched 98.26% DPPH radicals, as indicated in Figure 6.

On average, DPPH FRSAs of ethanol, methanol, and acetone extracts of the three cultivars of sweet potato were higher than 90%. In contrast, DPPH FRSA of ddH_2O and EA extracts of TNG66 and TNG 73 as well as the EA extract of TNG 57 were lower than 50% (Table 8). Diethyl ether and aqueous ethanol extracts showed intermediate effectiveness. Extracts of fresh roots of TNG57 were on average the most effective against DPPH free radical with an average DPPH FRSA value of 78.63%, while TNG66 and TNG73 had similar activities with DPPH FRSA values of 70.25 and 70.75%, respectively. The DPPH FRSA was higher in the skin sample extracts, followed by the unpeeled root, and then the peeled root ones (Table 8).

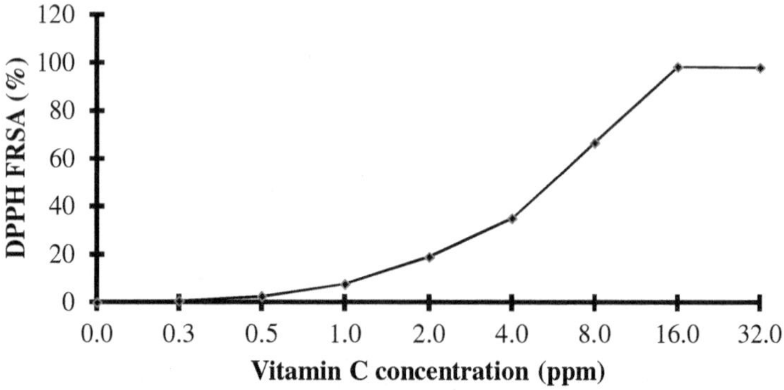

Figure 6. Dose-response curve of DPPH FRSA of vitamin C.

Table 8. Interactive effects of the solvent and genotype on the percentage DPPH FRSA of the extracts of TNG57, TNG66, and TNG73

Solvent[1]	(Mean±SE, n=3)[2]			
	TNG57	TNG66	TNG73	Mean/solvent
Ethanol	96.14±0.27^{a}	95.41±0.41^{a}	89.60±1.50^{bc}	93.71±0.77^{a}
Methanol	94.19±0.38^{a}	91.91±0.35^{b}	91.53±0.21^{b}	92.54±0.29^{a}
Acetone	85.73±5.09^{b}	88.53±3.63^{c}	97.84±0.13^{a}	90.70±2.25^{a}
Diethyl ether	74.76±7.82^{c}	88.96±0.76^{bc}	88.31±0.86^{c}	84.01±2.84^{b}
Aq. ETOH	84.09±2.68^{b}	66.77±3.61^{d}	60.97±2.53^{d}	70.61±2.54^{c}
Ethyl acetate	45.06±5.58^{e}	30.45±6.31^{e}	41.64±2.07^{e}	39.05±3.03^{e}
ddH$_2$O	70.43±3.00^{d}	29.77±10.26^{e}	25.35±8.35^{f}	41.85±5.89^{d}
Mean/cultivar	78.63±2.59^{A}	70.25±3.84^{B}	70.75±3.56^{B}	

[1]Aq. ETOH: aqueous ethanol; ddH$_2$O: distilled de-ionized water. [2]Means with same small letter in a column or capital letter in a row are not significantly different ($p>0.05$).

Significant interactions were observed among the three factors (solvent, genotype, and part of the root) influencing the DPPH FRSA. For instance, the difference between ethanol and methanol was not significant in term of DPPH FRSA of TNG57 and TNG73, while ethanol was significantly more effective than methanol for extracting TNG66 samples (Table 8). Methanol ranked as first best solvent before acetone for extracting the peeled root samples, while it ranked as third after acetone and ethanol for extracting the unpeeled root ones (Table 9). Higher DPPH FRSA were observed in peeled root extracts of TNG73 than in peeled root extracts of TNG66, while no significant difference was observed between the TNG66 and TNG73 in terms of DPPH FRSA of their unpeeled root. Moreover, the skin sample extracts of TNG66 had on average higher DPPH FRSA than those of TNG73 (Table 9).

4.3.3. Correlations between antioxidant activity (AOA) and total phenolic content

The regression parameters listed in Table 10 shows positive ($b>0$) and significant ($p<0.05$) pair-wise relationships between the DPPH FRSA and the total phenolics in the three cultivars. However, the analysis of the coefficient of determinations (R^2) shows that the contribution of total phenolics to the DPPH FRSA varied largely among the three cultivars (7.50, 25.81, and 51.19% in TNG57, TNG66, and TNG73, respectively). These results mean that total phenolics are major contributors to DPPH FRSA of TNG66 and TNG73, while TNG57 seems to have other important secondary metabolites involved in its AOA.

Table 9. Interactive effects of the solvent, part of the root, and genotype on the percentage DPPH free radical scavenging activity of the extracts of TNG57, TNG66, and TNG73

Solvent[1]	(Mean±SE, n=3)[2]			
	TNG57	TNG66	TNG73	Mean
Peeled root (PR)				
Methanol	95.39±0.03^{Aa}	91.52±0.87^{Aa}	91.90±0.05^{Aab}	92.93±0.67^{a}
Ethanol	95.31±0.04^{Aa}	93.99±0.39^{Aa}	89.26±0.07^{Ab}	92.85±0.93^{a}
Acetone	65.37±0.16^{Cc}	74.01±0.40^{Bd}	98.13±0.16^{Aa}	79.17±4.90^{b}
DE	43.54±0.43^{Cd}	86.45±0.12^{Bc}	90.53±0.03^{Aab}	73.50±7.52^{c}
Aq. ETOH	73.44±1.04^{Ab}	58.99±0.55^{Be}	55.50±0.26^{Bc}	62.64±2.77^{d}
EA	63.29±4.36^{Ac}	5.39±0.92^{Cf}	44.14±1.07^{Bd}	37.61±8.62^{e}
ddH_2O	60.99±0.73^{Ac}	8.40±4.83^{Bf}	7.08±2.54^{Be}	25.49±9.02^{f}
PR mean	71.05±3.95^{A}	59.82±7.90^{C}	68.07±7.02^{B}	66.31±3.75
Unpeeled root (UR)				
Methanol	94.39±0.08^{Aab}	92.87±0.00^{Aab}	92.01±0.03^{Aa}	91.60±0.31^{c}
Ethanol	97.06±0.08^{Aa}	96.42±0.13^{Aa}	94.96±0.14^{Aa}	92.14±1.90^{b}
Acetone	96.15±0.30^{Aa}	95.71±0.40^{Aa}	97.67±0.34^{Aa}	96.42±0.33^{a}
DE	92.01±0.32^{Ab}	91.65±0.40^{Ab}	89.46±0.06^{Ab}	87.49±0.64^{d}
Aq. ETOH	89.51±0.27^{Ab}	80.99±0.17^{Bc}	67.09±0.15^{Bc}	69.98±5.22^{e}
EA	42.66±0.34^{Ad}	43.92±0.39^{Ae}	37.28±1.51^{Ae}	38.26±3.65^{f}
ddH_2O	80.39±2.52^{Ac}	70.41±0.21^{Bd}	58.48±1.76^{Bd}	30.30±9.93^{g}
UR mean	80.24±5.10^{A}	69.23±6.85^{B}	67.47±6.59^{B}	72.31±3.61
Skin				
Methanol	92.78±0.05^{Aa}	91.34±0.08^{Aa}	90.69±0.03^{Aa}	93.09±0.35^{b}
Ethanol	96.03±0.27^{Aa}	95.82±0.48^{Aa}	84.57±0.04^{Aa}	96.15±0.32^{a}
Acetone	95.68±0.12^{Aa}	95.86±0.19^{Aa}	97.73±0.06^{Aa}	96.50±0.34^{a}
DE	88.74±0.17^{Aa}	88.77±0.06^{Aa}	84.96±0.18^{Aa}	91.04±0.43^{b}
Aq. ETOH	89.30±0.11^{Aa}	60.32±2.01^{Bb}	60.32±6.52^{Bb}	79.20±3.27^{c}
EA	29.25±7.73^{Bc}	42.03±2.20^{Ac}	43.51±5.82^{Ac}	41.29±1.12^{e}
ddH_2O	69.91±2.60^{Ab}	10.49±0.00^{Bb}	10.49±0.00^{Bd}	69.76±3.29^{d}
Skin mean	84.60±4.01^{A}	81.71±3.96^{B}	76.71±4.75^{C}	81.01±2.45

[1]DE: diethyl ether; Aq. ETOH: aqueous ethanol; EA: ethyl acetate; ddH_2O: distilled deionized water. [2]Means with a same capital letter in a row or small letter in a column within samples are not significantly different (p>0.05).

Table 10. Regression parameters between total phenolics (X) and DPPH FRSA (Y)

Cultivar	Equation (Y=a+bX)	Regression parameters[1]		
		p-value (Ho: b=0)	R^2	R
TNG57	Y=72.14+0.59X	0.0302	0.075	0.274
TNG66	Y=43.42+2.25X	0.0001	0.258	0.501
TNG73	Y=9.97+2.67X	0.0001	0.512	0.716

[1]Ho: null hypothesis; b=coefficient of regression; R^2: coefficient of determination; R: coefficient of correlation.

4.3.4. Heat stability of ethanol extracts of sweet potato antioxidant activity in liquid media incubated at increasing temperature

To analyze the effects of the incubation temperature on the DPPH FRSA, ethanol extracts were incubated with DPPH radical solution at various temperatures, from 25°C to 100°C for 30 min. Then the absorbance at 517 nm was read and the DPPH FRSA (%) determined. For TNG57, the DPPH FRSA value remained stable up to 75°C, and then it dropped to 89.11% at 100°C (Tables 11). For TNG66, the DPPH FRSA increased from 95.41% at 25°C to 97.67% at 50°C. Then, it dropped to 92.83% at 75°C and 88.50% at 100°C (Tables 12). For TNG73, the DPPH FRSA increased from 89.6% at 25°C to 96.3% at 50°C. Then, it dropped to 85.08% at 100°C (Tables 13). The pooeld results for the three cultivars amounted to 93.71±0.69% at 25°C, 97.22±0.15% at 50°C, 95.07±0.47% at 75°C, and 87.57±0.42% at 100°C. Then, on average, the DPPH FRSA slightly increased with the increase of the temperature up to 50 or 75°C, and then decreased.

Table 11. Percentage of DPPH FRSA of TNG57 extracts incubated at 25, 50, 75, and 100°C

Part[1]	(Mean±SE, n=3)[2]			
	25°C	50°C	75°C	100°C
PR	$95.31±0.04^{Aa}$	$97.05±0.18^{Aa}$	$96.58±0.30^{Aa}$	$88.22±0.01^{Ba}$
Skin	$97.06±0.27^{Aa}$	$98.91±0.08^{Aa}$	$98.55±0.24^{Aa}$	$89.90±0.07^{Ba}$
UR	$96.04±0.08^{Aa}$	$97.10±0.13^{Aa}$	$97.98±0.04^{Aa}$	$89.22±0.05^{Ba}$
Mean	$96.14±0.29^{A}$	$97.69±0.35^{A}$	$97.70±0.34^{A}$	$89.11±0.28^{B}$

[1]PR: peeled root, UR: unpeeled root. [2]Means with a same capital letter in a row or small letter in a column are not significantly different (p>0.05);

Table 12. Percentage of DPPH FRSA of TNG66 extracts incubated at 25, 50, 75, and 100°C

Part[1]	(Mean±SE, n=3)[2]			
	25°C	50°C	75°C	100°C
PR	93.99±0.39^{Ab}	94.95±0.07^{Ab}	95.58±0.17^{Aa}	87.21±0.05^{Bb}
Skin	96.42±0.48^{Ba}	99.07±0.15^{Aa}	97.15±0.20^{ABa}	89.95±0.06^{Ca}
UR	95.82±0.13^{Bab}	98.99±0.15^{Aa}	95.76±1.99^{Ba}	88.34±0.06^{Cab}
Mean	95.41±0.42^{B}	97.67±0.79^{A}	92.83±0.29^{C}	88.50±0.46^{D}

[1]PR: peeled root, UR: unpeeled root. 2Means with a same capital letter in a row or small letter in a column are not significantly different ($p>0.05$);

Table 13. Percentage of DPPH FRSA of TNG73 extracts incubated at 25, 50, 75, and 100°C

Part[1]	(Mean±SE, n=3)[2]			
	25°C	50°C	75°C	100°C
PR	89.26±0.07^{Bb}	95.75±0.08^{Aa}	94.23±0.20^{Aa}	81.96±0.05^{Cb}
Skin	84.57±0.14^{Cc}	97.10±0.26^{Aa}	95.48±0.10^{Aa}	89.60±0.02^{Ba}
UR	94.96±0.04^{Aa}	96.04±0.52^{Aa}	94.33±0.02^{Aa}	83.69±0.16^{Bb}
Mean	89.60±1.73^{B}	96.30±0.24^{A}	94.68±0.23^{A}	85.08±1.34^{C}

[1]PR: peeled root, UR: unpeeled root. [2]Means with a same capital letter in a row or small letter in a column are not significantly different ($p>0.05$);

4.4. Discussion

DPPH FRSAs of ethanol, methanol, and acetone extracts of the three cultivars of sweet potato were similar to results obtained with methanol extracts of some medicinal plants (Miliauskas *et al.*, 2004). AOA of extracts was solvent dependent due to different polarities of the different secondary metabolites involved in the anti-oxidative process (Moure *et al.*, 2001). In many cuisines, sweet potato skin is a by-product with little economic value; it was interesting to observe that the DPPH FRSA and the phenolic content were higher in the skin samples than in the flesh samples. Thus, sweet potato skin can constitute a non-expensive source of antioxidants for food industry to satisfy the general desire of the consumers to replace synthetic food additives with natural alternatives.

When extracted with ethanol, the phenolic contents of the extracts of fresh sweet potato TNG57, TNG66, and TNG73 corresponded to 153.15, 151.09, and 326.19 mg GAE/100 g fresh matter, respectively. These results showed a close relationship between the color of the samples and the phenolic contents, with higher value in red tissues. Total phenolic content of various flesh-colored sweet potatoes has been reported to vary from 0.3 to 95 mg GAE/100 g fresh matter with highest contents in the red flesh color cultivars (Teow *et al.*, 2007). The distribution pattern on basis of the flesh color was similar to the results of our study, and in light of those results, the three cultivars of sweet potato analyzed in this study can be considered a high phenolic content cultivars.

Genetic factors and growing conditions may play an important role in the formation of secondary metabolites, including phenolic compounds (Ooyanagui, 1984; Islam *et al.*, 2003). Similarly to the AOA, the total phenolic content was higher in the skin samples in comparison to the flesh, complying with Kondo *et al.* (2002) who reported higher total phenolic content in the skin than the flesh of fruits. Second and third order interrelations observed among part of the root, genotypes, and solvents suggested that the AOA and phenolic content of sweet potato root extracts are part of the root, genotype, and solvent-dependent. As a consequence, the results of solvent optimization obtained with one or some cultivars cannot be extrapolated to other cultivars and, solvent optimization should be a routine practice in laboratories that undertake AOA and secondary metabolites analyses.

The results of the heat stability test suggested that during the heating process in liquid media, the extracts' chemical composition changed, leading to higher AOA. However, when the heating temperature went beyond 50 or 75°C, the chemical changes led to a decrease in the DPPH FRSA of the extracts. Those results agreed with Larrauri *et al.* (1998) who observed reduction of AOA in red grape pomace peels dried with hot air at 100 °C. Rehman *et al.* (2003) reported that evaporation and decomposition at elevated temperatures results in loss of activity of synthetic antioxidants. For instance, α-tocopherol, is stable only up to 55°C restricting its application at high temperature. However, BHT activity toward DPPH scavenging remains unchanged at elevated temperatures (Ekanayake *et al.*, 2005).

4.5. Conclusions

The highest phenolic contents of TNG57 and TNG73 were observed in ethanol 95% followed by diethyl ether extracts, while TNG66 exhibited its highest total phenolic content in diethyl ether, followed by ethanol 95% extract. On average, ethanol, methanol, and acetone extracts showed the highest AOA, followed by diethyl ether extract. In addition to its high effectiveness to extract phenolic antioxidants, ethanol has the advantage of being a non-toxic solvent. Thus it was chosen for further extractions. On average, the AOA was higher in the skin than the flesh, and the DPPH FRSA was higher than 80% in the three cultivars when extracted with ethanol, methanol, acetone, or diethyl ether. TNG73 showed the highest average phenolic content, followed by TNG66 and TNG57. The carotenoid content may be considered as low for TNG73, intermediate for TNG57, and high for TNG66.

Chapter 5. Effect of freeze-drying on phenolic content and antioxidant activity of TNG57, TNG66, and TNG73 cultivars of sweet potato (*Ipomoea batatas* Lam.)

Abstract

We evaluated the effects of freeze-drying on total phenolics, total flavonoids, and antioxidant activity (AOA) of sweet potato. Flesh and skin samples of the cultivars TNG57, TNG66, and TNG73 were used. The AOA was assessed using 1, 1-diphenyl-2-picrylhydradzyl (DPPH) and 2,2'-Azino-di-[3-ethylbenzothiazoline-6-sulfonate] (ABTS) assays. Non-dried samples were used as control. The freeze-drying process resulted in decreases by 0.60 to 28% of total phenolics and 5.07 to 14.09% of total flavonoids. TNG73 exhibited the highest total phenolics and total flavonoids, followed by TNG66, and then TNG57. The total phenolics, total flavonoids, and AOA were consistently higher in the skin than the flesh samples. DPPH and ABTS free radical scavenging activities (FRSA) of both raw and freeze-dried samples exceeded 80% at extract concentration of 10 mg/ml and were concentration-dependent. The lowest IC_{50} values (highest AOA) were observed in TNG73, followed by TNG66, and then TNG57.

Key words: phenolic content, freeze-drying, sweet potato cultivar, DPPH assay, ABTS assay, free radical scavenging activity

5.1. Introduction

Reactive oxygen species (ROS) are often generated as byproducts of biological reactions (Lo *et al.*, 2004) and impair the biological activity of biomolecules such as DNA, lipids, and protein, causing diseases such as neurodegentrative disorders, cancer, and atherosclerosis (Aruoma, 1998). The human body in healthy individuals produces some antioxidants such as superoxide dismutase (SOD), glutathione peroxidase, and glutathione reductase, etc., which prevent the ROS damaging effects (Rimbach *et al.*, 2005). However, excessive production of free radicals can overpass the human defense mechanism and cause oxidative damage in cells. Studies have suggested that eating a diet rich in antioxidant-containing foods has been linked to a lower risk of the above mentioned diseases (Tonucci *et al.*, 1995).

Various reports mention the benefits of phenolic compounds in preventing oxidative damages. Scalbert *et al.* (2005) reported that phenolic compounds may play an important role in preventing chronic illnesses such as cardiovascular diseases, certain type of cancers, neurodegenerative diseases, and diabetes. Shimozono *et al.* (1996) reported that phenolic compounds are implicated in suppressing various health related disorders in mice, including melanogenesis. Phenolic compounds are also reported to prevent hepatoma invasion (Yagasaki *et al.*, 2000) and human immunodeficiency virus (HIV) replication (Zhu *et al.*, 1999). Suda *et al.* (1997) reported the ameliorative effect of juice extracted from red-fleshed sweet potato (*Ipomoea batatas* Lam.) against carbon tetrachloride induced liver injury in rats. According to Huang *et al.* (2004), sweet potato inhibits cancer cell proliferation, probably due to the synergistic effect of phenolics with other phytochemicals.

Sweet potato is an easy-to-grow crop with good adaptability to diverse environmental conditions, high yielding ability, and high energy content (Ravindran *et al.,* 1995). It is rich in dietary fiber, minerals, and vitamins (Woolfe, 1993). It can be grown with no synthetic pesticides and fertilizers (FAO, 1984). It is suitable for environmentally friendly agriculture and organic food production and is a promising crop for global food security (Scott, 1992). The AOA of phenolic compounds in sweet potato root tissue has been evaluated and reported to be higher than in many vegetables (Yamakawa and Yoshimoto, 2002). In terms of total phenolic content on a fresh matter basis, sweet potato ranked eighth among 23 most commonly consumed vegetables in the world (Vinson *et al.*, 1998). Thus, sweet potato used needs to be more diversified to extend its contribution to food security, disease prevention, and health promotion. As many consumers are looking for convenience, high quality, and healthy choice products, dried sweet potato could be considered as a high quality minimally processed product.

Minimally processed fruits and vegetables are popular value-added convenience foods widely available worldwide. Consumption of fresh-cut produce has grown at a rapid pace recently, and new product development is still occurring (Garrett, 2002). Minimally processed are now commercially available for direct consumer use or for further processing by restaurants and fast-food chains. Di Venere *et al.* (2005) reported an increase in total phenolic content and AOA in minimally processed carrots, potato, jicama roots, escarole, and radicchio. Like other vegetables, sweet potato can also be minimally processed and consumed without further cooking (Padda, 2006). Sweet potato cultivars TNG57, TNG66, and TNG73 were found to have high AOA, mainly due to a high total phenolic content (Duvivier *et al.*, 2008a). The objective of this study was to quantify the total phenolics, total flavonoids, and total AOA changes during freeze-drying of TNG57, TNG66, and TNG73 cultivars of sweet potato. TNG57, TNG66, and TNG73 sweet potato cultivars have been developed in Taiwan, and they have yellow, reddish yellow and red color, respectively, which makes them attractive to consumers (Duvivier *et al.*, 2008a).

Freeze-drying is usually considered as a mild and safe drying method (de Arantes-Frota *et al.*, 2008), but too expensive to scale up. Ratti (2001) reported that the energy efficiency of freeze dryers is notably low, with the sublimation of the water vapor consuming only 45% of the total energy, the other 55% of energy is used to freeze the product, pull a vacuum and condense moisture from the vacuum pumps. According to Grabowski (2002), freeze drying is the most expensive drying processes in regards to energy consumption and operation costs that are 4-8 times higher than conventional air-drying. However, freeze-drying can be used as a comparative research method in laboratory and the results can serve as a valuable reference for assessing other drying methods.

5.2. Materials and Methods

5.2.1. Materials

Freshly harvested sweet potato tuberous roots were purchased from a local traditional market in Pingtung, Taiwan, in August 2008. DPPH and ABTS radicals, catechin, ethanol, Folin-Ciocalteu phenol's reagent, gallic acid, hexahydrate aluminium trichloride, linoleic acid, sodium carbonate (Na_2CO_3), sodium hydroxide, sodium nitrite, thiobarbituric acid (TBA), trichloroacetic acid (TCA), and vitamin C were purchased from Sigma Chemical Co. (St. Louis, MO, USA). All the other reagents were either extra pure or analytical grade.

5.2.2. Experiment design

For total phenolic and total flavonoid analyses, the design consisted of a 3*2*2 factorial experiment (combination of three cultivars, three parts of the root, and two treatments: freeze-drying versus non-drying) in a completely randomized design with three replicates, yielding 12 treatments and 36 experiment units. For DPPH and ABTS assays, the design consisted of a 10*3*2*2 factorial experiment (combination of 10 extract concentrations, three cultivars, 3 parts of the root, and two treatments: freeze-drying versus non-drying) in a completely randomized design with three replicates, yielding 120 treatments and 360 experiment units.

5.2.3. Sample preparation and extraction

Sweet potato tuberous root samples were washed in tap water, left at room temperature for evacuation of the washing water, peeled, and cut in small pieces (0.5 to 1 cm of thickness). The sample dry mater was measured according to AOAC (1970). Both the flesh and skin samples were then divided into six parts. Three parts (three replicates) were frozen at -20°C for 24 h and freeze-dried at -40°C for 48 h in a laboratory freeze-dryer (model MCFD 5508, ILSHIN Lab Co., South Korea). The three other parts served as control and were not submitted to any drying process. The dried samples were ground into a fine powder using a laboratory miller.

The equivalent of 4 ±0.01 g dry matter of dried sample was mixed with 40 ml ethanol in a 250 ml flask. The equivalent of 4 ±0.01 g dry matter of fresh sample was blended with 40 ml ethanol using a high speed blender. The mixture was then transferred into a 250 ml flask. The flask with the mixture of dried or fresh sample and ethanol was then shaken at 100 rpm in darkness for 24 h in a water bath set at 25°C for complete extraction (Huang *et al.*, 2004). The mixture was filtered through Whatman no. 1 filter paper (Whatman Inc., Clifton, NJ). The residue was washed with an additional 40 ml extracting solvent. The filtrate was evaporated to dryness using a rotary evaporator model BUCHI 111 equipped with a water-bath BUCHI 461 (Switzerland) and a water vacuum pump EYELA A-3S (Tokyo, Japan). The dried residue, referred to as crude extract, was weighed, re-dissolved in ethanol to a concentration of 10 mg/ml, and stored at -20 °C until analysis. The extraction yield (%) was estimated dividing the dry extract weight by the sample dry matter and multiplying by 100.

5.2.4. Total phenolic and total flavonoids assays

The total phenolic assay was conducted according to the method of Folin and Ciocalteu (1927) as reported by Miliauskas *et al.* (2004) using gallic acid as standard. The results were reported in mg gallic acid equivalent/100g dry matter. The total flavonoids content was assessed by the aluminum trichloride colorimetric method (Marinova *et al.*, 2005) using catechin as standard. The results were reported in mg catechin equivalent (CE)/100g dry matter.

5.2.5. Antioxidant activity (AOA) assays

Ten different extract concentrations ranging from 0.02 to 10 mg/ml were prepared and tested for total AOA using DPPH and ABTS assays. DPPH assay was carried out as described in Molyneux (2004). A 100 μM DPPH stock solution was prepared by dissolving 4 mg DPPH radical in 100 ml of pure ethanol. Then, 1.5 ml of stock solution was added to 4.5 ml of 0.02 to 10 mg/ml plant extract and vortex-mixed. Separately, 1.5 ml of 100 μM DPPH solution was mixed with 4.5 ml ethanol resulting in a 6 ml blank. After 30 min incubation in dark at room temperature, absorbencies of the test sample and blank were read at 517 nm in three replicates. Total AOA of the samples as DPPH FRSA was determined using the following formula: DPPH FRSA (%) $=[(A_0-A_f)/A_0]*100$, where A_0 is the blank absorbance at 517 nm after 30 min incubation and A_f, test sample absorbance at 517 nm after 30 min incubation.

ABTS assay was performed as reported by Teow *et al.* (2007). An ABTS radical solution containing 8 mM ABTS salt and 3 mM potassium per-sulfate/l in ddH_2O was prepared. The solution was kept at room temperature in the dark for 16 h before use. The ABTS radical solution (2 ml) was diluted 67-fold with 95% ethanol (to 134 ml) in order to obtain an absorbance between 0.8 and 0.9 at 734 nm. Fresh ABTS radical solution was prepared for each analysis. An aliquot of 20 μl plant extract (0.02 to 10 mg/ml) or 95% ethanol (as blank) were mixed with 980 μl diluted ABTS radical solution. Total AOA of the samples as ABTS FRSA was determined using the following formula: ABTS FRSA (%) $=[(A_0-A_f)/A_0]*100$, where A_0 is the blank absorbance at 734 nm after 30 min incubation and A_f = test sample absorbance at 734 nm after 30 min incubation.

The data of AOA values were plotted against the extract concentrations, the IC_{50} value, that is the concentration (mg/ml) that produces 50% free radical inhibition, was computed by graphic interpolation according to Stock (1939) for each replicate, and a mean IC_{50} value was computed for each treatment. Treatments with lower IC_{50} values were considered to have higher AOA. A plot of AOA versus extract concentration was reported for each treatment.

5.2.6. Statistical analysis

The data were submitted to analysis of variance according to Kuehl (1999). The main factors' effects and interactions were determined by Fisher-Snedecor's F test at $\alpha=0.05$. Results were reported as mean ± SE. Means were compared by Duncan multiple range test. The SAS (Statistical Analysis System, v. 8.1; SAS Institute Inc., Cary, NC, USA) software was used.

5.3. Results

5.3.1. Dry matter content and extraction yield of the samples

For the raw samples, the dry matter content ranged from 26.4 to 30.69 % as indicated in Table 14. On average for raw samples, TNG66 had higher dry matter content, followed by TNG57, and then TNG73. For freeze-dried samples, the lowest dry matter content was observed in TNG66 flesh (94.08 %), while the highest was in TNG57 skin (99.70 %). The extraction yield was consistently higher in the raw samples (9.79 to 27.27 %) than the freeze-dried ones (3.44 to 13.18 %). On average, TNG73 had the highest extraction yield, followed by TNG66, and then TNG57 for both raw and freeze-dried samples.

5.3.2. Total phenolic and total flavonoid contents

The total phenolic content varied from 56.78 to 1769.30 mg GAE/100 g sample dry matter in raw samples. In freeze-dried samples, it varied from 48.21 to 1758.22 mg GAE/100 g sample dry matter (Table 15). Both for raw and freeze-dried samples, the highest content of total phenolics were observed in TNG73, followed by TNG66, and then TNG57. The skin had consistently higher phenolic content than the flesh. The freeze-drying process resulted in decreases of the total phenolic content from 0.60 % in the skin of TNG73 to 28 % in the flesh of TNG66 (Figure 7). On average, higher decreases were observed in the flesh in comparison to the skin. The highest decreases of the total phenolics were observed in TNG66, followed by TNG57, and then TNG73.

The total flavonoid content showed a quite similar trend. It varied from 39.59 to 711.53 mg CE/100 g sample dry matter in raw samples, and from 35.47 to 670.47 mg CE/100 g sample dry matter in freeze-dried samples (Table 15). The freeze-drying process resulted in decreases of total flavonoids from 5.07 % in the skin of TNG73 to 14.9 % in the flesh of TNG66 (Figure 7). Higher decreases of total flavonoids were consistently observed in the flesh in comparison to the skin, and the highest decreases were observed in TNG66.

Table 14. Dry matter content and extraction yield of raw and freeze-dried samples of TNG57, TNG66, and TNG73

Cultivar	Part of the root	(Mean±SE, n=3)[1]	
		Dry matter (%)	Extraction yield (% dry matter)
Control			
TNG57	Flesh	27.00 ± 0.50^{Bb}	9.79 ± 0.07^{Ad}
	Skin	28.00 ± 0.60^{Bb}	10.23 ± 0.33^{Ad}
TNG66	Flesh	30.69 ± 0.15^{Ba}	19.98 ± 0.22^{Ab}
	Skin	30.52 ± 0.20^{Ba}	6.33 ± 0.06^{Ae}
TNG73	Flesh	26.40 ± 0.28^{Bc}	12.82 ± 0.37^{Ac}
	Skin	26.70 ± 0.30^{Bc}	27.27 ± 0.84^{Aa}
Freeze-drying			
TNG57	Flesh	97.35 ± 0.35^{Ab}	3.87 ± 0.30^{Bd}
	Skin	99.70 ± 0.30^{Aa}	3.73 ± 0.31^{Bd}
TNG66	Flesh	94.08 ± 0.29^{Ac}	3.44 ± 0.29^{Bd}
	Skin	98.18 ± 0.28^{Ab}	5.02 ± 0.50^{Bc}
TNG73	Flesh	98.23 ± 0.31^{Ab}	8.95 ± 0.95^{Bb}
	Skin	97.45 ± 0.30^{Ab}	13.18 ± 1.03^{Ba}

[1]Means with different capital letters in a column across treatment (freeze-drying or control) are significantly different ($p<0.05$); means with different small letters in a column within treatment (freeze-drying or control) are significantly different ($p<0.05$).

Table 15. Content of total phenolics and total flavonoids of raw and freeze-dried sweet potato cv TNG57, TNG66, and TNG73

Cultivar	Part	(Mean±SE, n=3)[1]	
		Total phenolics[2]	Total flavonoids[3]
Control			
TNG57	Flesh	56.78 ± 1.45^{Af}	39.59 ± 0.00^{Af}
	Skin	169.43 ± 2.21^{Ae}	68.49 ± 0.51^{Ae}
TNG66	Flesh	223.25 ± 3.76^{Ad}	124.27 ± 1.67^{Ad}
	Skin	235.86 ± 3.86^{Ac}	124.88 ± 4.85^{Ac}
TNG73	Flesh	532.58 ± 3.17^{Ab}	255.31 ± 2.12^{Ab}
	Skin	1769.30 ± 6.03^{Aa}	711.53 ± 2.95^{Aa}
Freeze-drying			
TNG57	Flesh	48.21 ± 1.56^{Be}	35.47 ± 0.05^{Bf}
	Skin	163.04 ± 2.03^{Bd}	64.88 ± 0.10^{Be}
TNG66	Flesh	160.56 ± 1.95^{Bd}	105.75 ± 0.23^{Bd}
	Skin	172.29 ± 2.50^{Bc}	112.79 ± 0.50^{Bc}
TNG73	Flesh	515.54 ± 4.27^{Bb}	232.67 ± 2.01^{Bb}
	Skin	1758.22 ± 6.11^{Ba}	675.47 ± 4.70^{Ba}

[1]Means with different capital letters in a column across treatment (freeze-drying or control) are significantly different ($p<0.05$); means with different small letters in a column within treatment (freeze-drying or control) are significantly different ($p<0.05$). [2]Results in mg gallic acid equivalent/100 g dry matter. [3]Results in mg catechin equivalent/100 g dry matter.

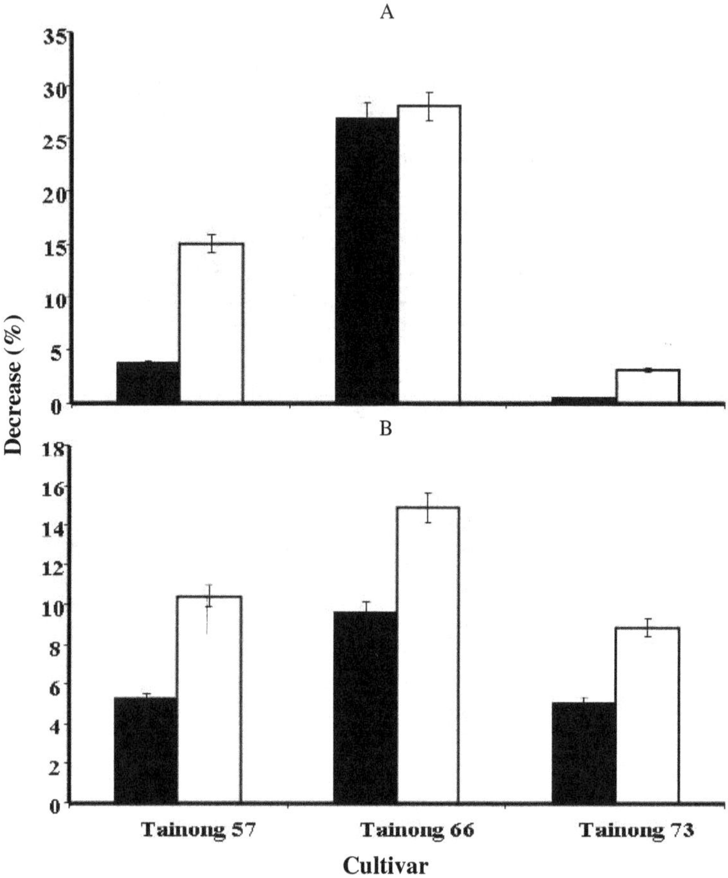

Figure 7. Decrease of the total phenolics and total flavonoids of TNG73 during freeze-drying.

Results are mean±SE of 3 replicates; A: total phenolics, B: total flavonoids, □: flesh, ■: skin.

5.3.2. DPPH free radical scavenging activity

The DPPH FRSA of the freeze-dried and raw samples summarized in Figure 8 show that the AOA of the sweet potato extracts was dose-dependant. Both the flesh and skin samples of the three cultivars exhibited high AOA (DPPH inhibition > 80%) at high concentrations of 5 and 10 mg/ml. At such high concentrations, no significant differences were observed between the flesh and the skin of the three cultivars. However, at lower concentrations, the different cultivars behaved differently.

For the raw samples, the skin of TNG57 exhibited higher DPPH inhibition than the flesh at concentrations of 0.02 to 2.5 mg/ml, while the skin samples of TNG66 showed higher DPPH FRSA than the flesh at concentrations of 0.31 to 2.5 mg/ml. The skin of TNG73 showed higher DPPH FRSA than the flesh at concentration of 0.16 mg/ml. For the freeze-dried samples, the skin of TNG57 showed higher AOA then the flesh at concentrations of 0.16 to 2.5 mg/ml, while the skin of TNG66 showed higher AOA than the flesh at concentrations of 1.25 and 2.50 mg/ml. For TNG73, differences were observed between the skin and the flesh at concentrations of lower and equal to 0.31 mg/ml, showing higher activities for the skin samples.

The DPPH IC_{50} of the extracts was computed to ease the evaluation of the freeze-drying treatment effect on the AOA. The computation DPPH IC_{50} was important because of the interaction of the extract concentration with the part of the root (flesh or skin) and the genotype on the DPPH FRSA. The results summarized in Table 16 show that the flesh of TNG73 had consistently the lowest DPPH IC_{50} values for both freeze-drying treatment and the control. It was followed in ascending order by TNG66, and then TNG57. The control skin samples showed a similar trend.

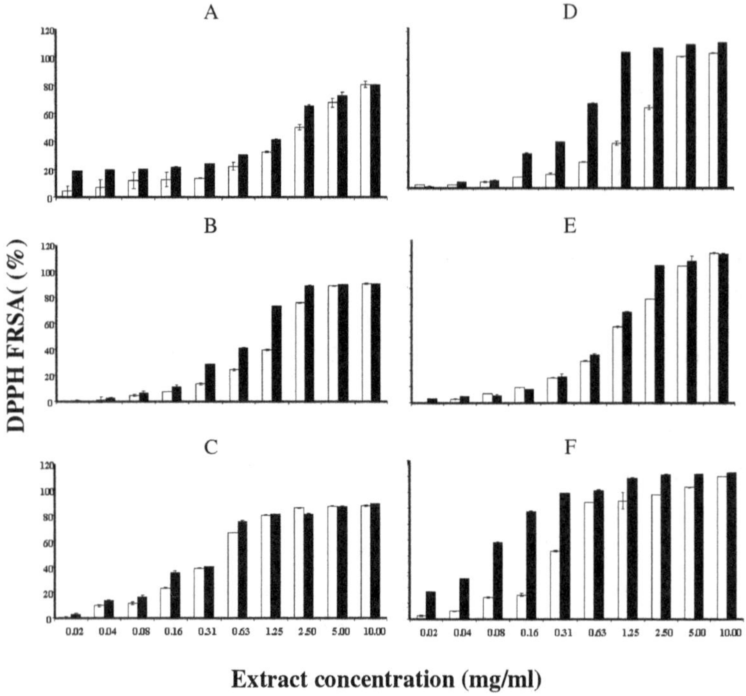

Figure 8. DPPH FRSA of of raw and freeze-dried sweet potato cv TNG57 TNG66, and TNG73.

A, B, C: raw samples of TNG57, TNG66, and TNG73, respectively; D, E, F: freeze-dried samples of TNG57, TNG66, and TNG73, respectively; □: flesh, ■ : skin; results are mean±SE of 3 replicates.

Table 16. DPPH IC_{50} (mg/ml) of raw and freeze-dried sweet potato cv TNG57, TNG66, and TNG73

Cultivar	Part	(Mean±SE, n=3)[1]	
		Control	Freeze-drying
TNG57	Flesh	2.50 ± 0.02^{Aa}	2.48 ± 0.02^{Aa}
	Skin	1.70 ± 0.38^{Ab}	0.61 ± 0.29^{Bc}
TNG66	Flesh	1.61 ± 0.11^{Ab}	1.52 ± 0.04^{Ab}
	Skin	0.79 ± 0.02^{Ac}	1.12 ± 0.06^{Ab}
TNG73	Flesh	0.16 ± 0.03^{Ad}	0.38 ± 0.01^{Ac}
	Skin	0.40 ± 0.08^{Ad}	0.35 ± 0.00^{Ac}

[1]Means with different capital letters in rows or small letters in columns are significantly different ($p<0.05$).

For the freeze-dried skin samples, TNG66 exhibited the highest DPPH IC_{50} values, followed by TNG57 and TNG73. On average, TNG73 showed the lowest DPPH IC_{50} value (0.32±0.03 mg/ml), followed by TNG66 (1.26±0.10 mg/ml), and then TNG57 (1.83±0.25 mg/ml). The average DPPH IC_{50} value was lower in the skin (0.82±0.16 mg/ml) than the flesh (1.44±0.27 mg/ml). The freeze-drying treatment with an average value of 1.19±0.25 did not result in significant decrease of the DPPH IC_{50} in comparison to the raw samples (1.08±0.23 mg/ml). The results of total phenolics and total flavonoids correlated negatively ($R<0$) and significantly ($p<0.05$) with the DPPH IC_{50}.

5.3.2. ABTS free radical scavenging activity

The ABTS FRSA of the raw and freeze-dried samples summarized in Figure 9 show that the AOA of the sweet potato extracts was dose-dependant. Both the flesh and skin samples of the three cultivars exhibited high AOA (ABTS inhibition > 80%) at high concentrations of 10 mg/ml. Similarly to the DPPH assay, at such high concentration, no significant differences were observed between the flesh and the skin. Skin samples of both raw and freeze-dried TNG66 exhibited ABTS FRSA higher than 80% at extract concentrations higher or equal to 5 mg/ml, while skin samples of both raw and freeze-dried TNG73 showed ABTS FRSA higher than 90% at concentration higher or equal to 2.5 mg/ml. The skin samples showed higher ABTS FRSA than the flesh at concentration of 5 mg/ml in both raw and freeze-dried TNG66, concentrations of 0.63 to 5 mg/ml for freeze-dried TNG73, and concentration of 2.5 mg/ml for raw TNG73. The ABTS IC_{50} of the extracts was computed to ease the evaluation of the freeze-drying treatment effect on the AOA.

The computation ABTS IC_{50} was important because of the interactions of the extract concentration with the part of the root (flesh or skin) and the genotype on the ABTS FRSA.

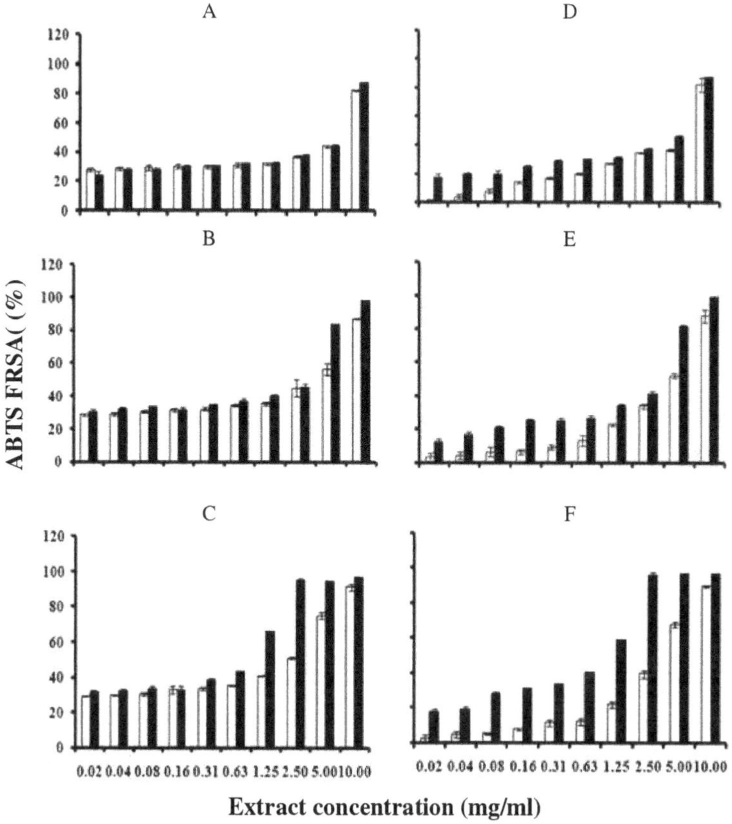

Figure 9. ABTS FRSA of raw and freeze-dried sweet potato cv TNG57, TNG66, and TNG73.

A, B, C: raw samples of TNG57, TNG66, and TNG73, respectively; D, E, F: freeze-dried samples of TNG57, TNG66, and TNG73, respectively; □: flesh, n : skin; results are mean±SE of 3 replicates.

The results summarized in Table 17 show that TNG73 consistently had the lowest ABTS IC_{50} values, followed by TNG66, and then TNG57. The skin had also lower ABTS IC_{50} than the flesh. The freeze-drying process of the flesh samples resulted in significant increases of the ABTS IC_{50} value from 14.62 to 43.88%. No significant increase of the ABTS IC_{50} value was observed for the skin samples. The average ABTS IC_{50} value was lower for the skin (3.12.±0.46 mg/ml) than the flesh samples (4.38±0.34 mg/ml). The freeze-drying treatment with an average value of 3.98±0.73 mg/ml resulted in a decrease of the ABTS IC_{50} of 13.39% in comparison to the raw samples (3.51±0.42 mg/ml).

The results of total phenolics and total flavonoids correlated negatively ($R<0$) and significantly ($p<0.05$) with the ABTS IC_{50} (Table 18). A moderate correlation was also observed between DPPH and ABTS assays.

Table 17. ABTS IC_{50} of raw and freeze-dried sweet potato cv TNG57, TNG66, and TNG73

Cultivar	Part of the root	(Mean±SE, n=3)[1]	
		Control	Freeze-drying
TNG57	Flesh	5.67±0.02^{Ba}	6.50±0.29^{Aa}
	Skin	5.67±0.03^{Aa}	5.33±0.44^{Ab}
TNG66	Flesh	3.65±0.02^{Bb}	4.69±0.05^{Ac}
	Skin	2.78±0.02^{Ac}	3.02±0.15^{Ad}
TNG73	Flesh	2.37±0.05^{Bc}	3.41±0.23^{Ad}
	Skin	0.94±0.03^{Ad}	0.95±0.03^{Ae}

[1]Means with different capital letters in rows or small letters in columns are significantly different ($p<0.05$).

Table 18. Pearson correlation coefficients (R) of total phenolics and total flavonoids with the antioxidant activities

Variables	Statistics	Total phenolics	Total flavonoids	DPPH IC_{50}
DPPH IC_{50}	R	-0.601*	-0.620*	
	p-value	0.039	0.032	
ABTS IC_{50}	R	-0.813**	-0.844**	0.784*
	p-value	0.001	0.001	0.003

*Significant correlation ($p<0.05$); **highly significant correlation ($p<0.01$).

5.4. Discussion

The total phenolics, total flavonoids, and AOA, of the skin samples were consistently higher than the flesh, complying with Kondo *et al.* (2002) who reported total higher phenolics content in the skin than the flesh of fruits. Those results also showed that sweet potato skin, which is considered as a waste in many cuisines, can constitute a valuable source of antioxidants. As dietary habit is part of the culture and difficult to change, maybe even research results show that sweet potato skin is rich in antioxidants, some cuisines will still consider it as a waste and discard it. Thus the food industry can exploit it as a non-expensive source of antioxidants to be used in alternative to the traditional synthetic additives. The highest content of phenolics was observed in TNG73 (red flesh cultivar), followed by TNG66 (reddish yellow cultivar), and then TNG57 (yellow cultivar). Those results were in agreement with Teow *et al.* (2007). Significant correlations of total phenolics and total flavonoids were observed with DPPH IC_{50}. Those results agreed with Teow *et al.* (2007) and Kurata *et al.* (2007).

Decreases from 0.60 to 28% of the total phenolic content were observed during the drying process. However, no significant decrease of the DPPH FRSA was observed during the freeze-drying process. With regards to information reported in the literature on losses of AOA during heat drying, those results suggested that freeze-drying is an effective method to dry sweet potato in order to keep acceptable levels of phenolics and AOA. For instance, Larraruri *et al.* (1998) observed reduction of AOA by 28% in the red grape pomace peels when dried with hot air at 100°C. Muchoki *et al.* (2007) observed that heating and drying of vegetables lead to substantial loss of antioxidants. Rehman *et al.* (2003) reported that evaporation and decomposition at elevated temperatures results in loss of AOA. Nonetheless, freeze-drying is an expensive technology, and thus, hardly applicable in developing and low-income countries.

5.5. Conclusions

The total phenolic content of the raw samples varied from 56.78 mg GAE/100g dry matter to 1769.30 GAE/100 g dry matter, while for the freeze-dried samples, it varied from 48.21 mg GAE /100 g dry matter to 1758.22 mg GAE/100g dry matter in the skin of TNG73. These changes represented decreases from 0.60 to 28% during the freeze-drying process. The total flavonoids content showed quite similar changes with decreases from 14.32 to 18.46%. The AOA of both raw and freeze-dried samples of the three cultivars were high (>80% inhibition in both DPPH and ABTS assays) and dose-dependant. The freeze-drying process resulted in decreases from 14.32 to 43.88% of the ABTS FRSA. Those changes were low compared to those caused by hot air drying reported in literature. The skin samples exhibited consistently higher phenolic content and AOA the flesh. The highest phenolic content and AOA were observed in TNG73, followed by TNG66, and then TNG57.

Chapter 6. Effects of different air drying methods on color parameters, antioxidant activity, phenolic, and carotenoid content of TNG57, TNG66, and TNG73 sweet potato (*Ipomoea batatas* Lam.) cultivars

Abstract

The air-drying effects on antioxidant activity (AOA), phenolic and carotenoid contents of sweet potato (*Ipomoea batatas* Lam.) cultivars TNG57, TNG66, and TNG73 were analyzed. Samples were submitted to sun-drying, low-temperature drying (LTD) at 25°C, hot air drying (HAD) at 50°C, or HAD at 75°C, extracted with ethanol, and analyzed for the above-mentioned parameters. Freeze-drying was used as control treatment. AOA was assessed by 1, 1-diphenyl-2-picrylhydradzyl (DPPH); anti-oxidative potency in linoleic acid system model (AOP), and 2, 2'-Azino-di-[3-ethylbenzothiazoline-6-sulfonate] (ABTS) assays. The results showed that all air-drying processes resulted in losses of AOA, phenolic and carotenoid contents with regards to the freeze-drying. Increased drying temperatures were accompanied with increased losses of AOA, phenolics, and carotenoids for TNG57 and TNG66. However, for TNG73, the losses of AOA, phenolics, and carotenoids were lower for samples dried at higher temperature for shorter time.

Key words: antioxidant activity, carotenoids, drying temperature, drying time, phenolics

6.1. Introduction

Sweet potato is an easy-to-grow crop with good adaptability to diverse environmental conditions, high yielding ability and high energy content (Ravindran *et al.*, 1995). It is suitable for organic food production and other environmentally friendly agricultural practices. It ranks fifth most important food crop in terms of fresh matter after rice, wheat, corn, and cassava in developing countries where 95% of its annual production (more than 133 million tons) is concentrated (International Center of Potato (CIP), 1998). Some cultivars have anti-carcinogenic properties and/or AOA (Teow *et al.*, 2007). Thus, it is a promising crop for global food security (Scott, 1992).

Food storage is an important factor of food and nutrition security and food drying by heat is an old and common practice of preserving food. Drying brings food water activity to a level unfavorable for the microorganisms' development during storage and facilitates food distribution by reducing bulk volume. However, thermal processing has long been perceived to cause the loss of some heat-labile nutrients, thus lowers the nutritional value of the food. Earlier studies in selected legume sprouts and seedlings indicate that thermal processing significantly alters functionality of legumes (Randhir and Shetty, 2005). The suitability of a post harvest food treatment depends on its effect on nutritious and functional properties of the food (Marc *et al.*, 2004).

Previous experiment showed that freeze-drying is an effective method to dry sweet potato in order to keep the phenolic content and the AOA. However, freeze-drying is an expensive technology, and thus, difficult to apply, particularly, in low-income developing countries. Effective and less expensive drying technologies are to be found to produce non-expensive food commodities. The objective of this step of my study was to analyze the effects of four drying processes: sun-drying, low-temperature drying (LTD) at 25°C, hot air drying (HAD) at 50°C, and HAD at 75°C, on the color parameters, AOA, total phenolic, and carotenoid contents of TNG73 sweet potato. Due to its proven effectiveness, freeze-drying was used as control.

6.2. Materials and Methods

6.2.1. Materials

Freshly harvested sweet potato tuberous roots were purchased from a local market in Pingtung, Taiwan, in September 2007. The samples were processed with not storage after purchase. The radicals DPPH and ABTS, catechin, Folin-Ciocalteu phenol's reagent, gallic acid, hexahydrate aluminium trichloride, linoleic acid, sodium carbonate (Na_2CO_3), sodium hydroxide, sodium nitrite, thiobarbituric acid (TBA), trichloroacetic acid (TCA), and vitamin C were purchased from Sigma Chemical Co. (St. Louis, MO, USA). All the other reagents were either extra pure or analytical grade.

6.2.2. Experiment design

Completely randomized design (CRD) with three replicates was used throughout the study. Each air-drying process (freeze-drying, sun-drying, LTD, HAD at 50°C, or HAD at 75°C) was a treatment, and freeze-drying was used as control. The response variables were DPPH FRSA, AOP, ABTS FRSA, RDI, total phenolic, flavonoid, carotenoid, and beta-carotene contents.

6.2.3. Sample preparation and color parameter measurements

Sweet potato tuberous root samples were washed in tap water, dried at room temperature, peeled, and cut in small pieces (0.5 to 1 cm of thickness). The sample dry mater was measured according to AOAC (1970). Water activity was measured in an Aw Quick water activity meter (Rotronic Instrument Corp., Huntington, N.Y., U.S.A.). The samples were then divided into five parts and submitted to freeze-drying, open sun-drying, LTD, or HAD in three replicates. LTD was done at 25 ^{0}C in a close-type low-temperature dryer adjustable between the temperature range of 0 to 50^{0}C. HAD was done at 50°C or 75°C in a conventional open type drying oven. A drying curve was established in each drying process (except freeze-drying) to decide the end-point drying time. This end-point was set at the time the drying curve stabilized. The dried samples were ground into a fine powder using a laboratory miller. The powder was weighed and sealed in laminated plastic bag and kept in desiccators until extraction. Figure 10 illustrates the sample preparation process. The color parameters were evaluated as described in section 3.2.3.

6.2.4. Crude extract preparation

An aliquot of 4 ±0.01 g sample was mixed with 40 ml ethanol in a 250 ml flask and extracted as described in section 4.2.3 for AOA and phenolic analyses. After filtration, the residue was washed with an additional 40 ml ethanol. The extraction yield (%) was estimated dividing the dry extract weight by the sample dry matter and multiplying by 100.

6.2.5. Carotenoids extraction

Carotenoids were extracted according to Teow *et al.* (2007). Briefly, an aliquot of 5 g powder were mixed with 2 g Na_2CO_3, 1 g diatomaceous earth, and 25 ml methanol. A hexane–acetone (1:1 v/v) mixture (50 ml) was added and stirred for 2 h in the dark. The mixture was filtered through Whatman no. 1 filter paper. The residue was first washed twice with 25 ml methanol, then once with 50 ml hexane–acetone mixture. The extract was combined in a 250 ml separatory funnel and completed to volume with ddH_2O. The aqueous phase was discarded and the upper layer was transferred into a 50 ml volumetric flask and made to volume (50 ml) with hexane for analysis (Chandler and Schwartz, 1998).

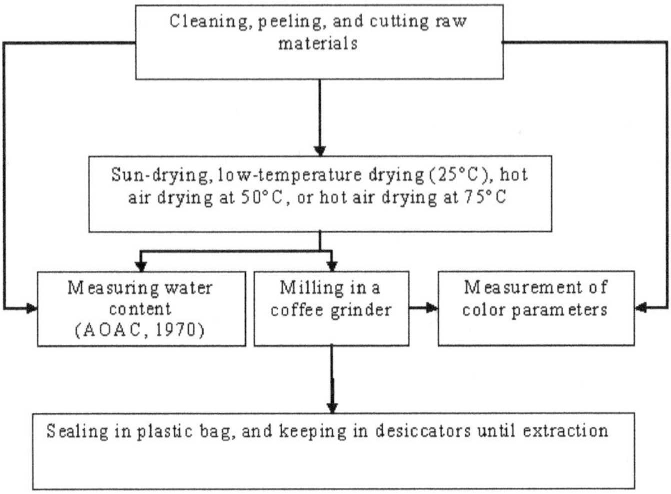

Figure 10. Flowchart of samples preparation to assess the effects of the drying process on the antioxidant activity, phenolic, and carotenoid contents.

6.2.6. Total phenolic, total flavonoid, total carotenoid, and beta-carotene contents assays

The total phenolic content was assayed as described in section 3.2.5. The total flavonoids content was assessed by the aluminum trichloride colorimetric method (Miliauskaus *et al.*, 2004; Marinova *et al.*, 2005) using catechin as standard. An aliquot (1 ml) ethanolic catechin standard solutions (20, 40, 60, 80, 100 mg/l) or plant extract (10 g/l) was mixed with 4 ml ddH_2O, 0.3 ml 5% $NaNO_2$ in ddH_2O, 0.3 ml 10% $AlCl_3$-6H_2O in ddH_2O, and 2 ml 1 M NaOH. The mixture was completed to volume (10 ml) with ddH_2O, vortex-mixed and incubated at 20°C in darkness for 40 min. Absorbance was read in triplicate at 510 nm after 40 min incubation at 20°C in darkness. We drew a standard catechin calibration curve, y=aX+b, where Y is the pure catechin content in mg/l sample analyzed; a, the slop of the regression; X, the sample absorbance at 510 nm (A_{510}); and b, the intercept of the equation. The standard catechin calibration curve was used to calculate the amount of flavonoids in plant extracts as mg catechin equivalent (CE)/l plant extracts (z) replacing X by A_{510} in the above equation. The results were then converted into mg/100 g dry matter. Total carotenoid and beta-carotene contents were analyzed as described in section 3.2.6.

6.2.7. Determination of the antioxidant activity

The AOA was assessed using DPPH, AOP, and ABTS assays. DPPH assay was conducted as described by Molyneux (2004). A 100 μM DPPH stock solution was prepared by dissolving 4 mg DPPH radical in 100 ml of pure ethanol. Then, 1.5 ml of stock solution was added to 4.5 ml of 5 mg/ml plant extract and vortex-mixed. Separately, 1.5 ml of 100 μM DPPH solution was mixed with 4.5 ml ethanol resulting in a 6 ml blank. After 30 min incubation in dark at room temperature, absorbencies of the test sample and blank were read at 517 nm in three replicates. Total AOA of the samples as DPPH FRSA was determined using the following formula: DPPH FRSA (%) =$[(A_0-A_f)/A_0]*100$, where A_0 is the blank absorbance at 517 nm after 30 min incubation and A_f, test sample absorbance at 517 nm after 30 min incubation.

The AOP was assessed according to a procedure described in Jung *et al.* (2005). 4 ml ethanolic plant extract (1 mg/ml) were mixed with 4 ml linoleic acid (2.5% v/v) dissolved in 99.5% ethanol, 8 ml phosphate buffer (0.05 M, pH 7.0), and 4 ml ddH_2O. The mixture was kept in a screw-cap container in darkness at 40°C for 78 h. This solution (1 ml) was added to a mixture (1 ml) containing 20% TCA and 0.75% TBA in methanol and then placed in boiling water for 30 min. After cooling, n-butanol (2 ml) was added and the mixture was shaken vigorously and centrifuged at 3000 rpm for 5 min at 5°C. The organic (upper) layer was taken, and its absorbance at 532 nm was measured against blank. The AOP (%) was calculated as follow: AOP (%) =[1-($A_{532\ sample}$/$A_{532\ blank}$)*100], where A_{532} is the absorbance at 532 nm.

The ABTS FRSA was determined by a procedure of Miller and Rice-Evans (1997) as reported in Teow *et al.* (2007). An ABTS radical solution containing 8 mM ABTS salt and 3 mM potassium per-sulfate/l in ddH_2O was prepared. The solution was kept at room temperature in the dark for 16 h before use. A 2 ml ABTS radical solution was diluted 67-fold with 95% ethanol in order to obtain an absorbance between 0.8 and 0.9 at 734 nm. Fresh ABTS radical solution was prepared for each analysis. A 20 μl plant extract (5 mg/ml) or 95% ethanol (as blank) was mixed with 980 μl diluted ABTS radical solution.

Total AOA of the samples as ABTS FRSA was determined using the following formula: ABTS FRSA (%) =[(A_0-A_f)/A_0]*100, where A_0 is the blank absorbance at 734 nm after 30 min incubation and A_f = test sample absorbance at 734 nm after 30 min incubation. Vitamin C was used as standard (Bouayed *et al.*, 2007). Aliquots of 20 μl Vitamin C standard solutions (50 to 1500 ppm) were separately mixed with 980 μl diluted ABTS radical solution in order to obtain final concentrations of 1 to 30 ppm. The absorbance at 734 nm was read after 30 min incubation in dark at 30°C. A vitamin C standard calibration curve was drawn and used to compute the ABTS FRSA expressed as vitamin mg C equivalent (VCE)/ 100 g dry matter.

6.2.8. Determination of a single antioxidant activity indicator

Many methods for evaluation of total AOA exist and no correlation has been established among them. Thus, no absolute ranking has been done for them. As yet, the choice of one of the different existing methods is done in function of the available resources, the analysis to be done, and its cost (Arnao *et al.*, 1999). No absolute superiority of one in comparison to other has been established. This lack of correlation and hierarchy among the different methods of analysis has posed a serious problem in ranking food crops on basis of AOA. In this study, to overcome this problem, a single AOA indicator was computed from the results of the three initial indicators (DPPH FRSA, AOP, and ABTS FRSA). It was referred to as AOA relative desirability index (RDI) and corresponded to an integrated criterion of ranking on basis of the AOA evaluated not only by one, but by three variables. The Gauss-Laplace normalization formula (Kuehl, 1999), slightly modified (Duvivier *et al.*, 2008b), was used to compute RDI.

$$RDI = \frac{\sum_{i=1}^{i=k} \frac{X_{it} - \overline{X_i}}{\sigma_i}}{k} + C.$$

In this formula, for a given AOA assay i (ABTS, AOP, or DPPH assay) and the corresponding response variable X_i, X_{it} is the value of X_i for the drying method t, while $\overline{X_i}$ and σ_i are respectively mean and standard deviation for all sample treatments, that is, for all X_{it}. K is the number of AOA assays, whereas C is a constant that equates the lowest RDI value to 1. The treatment with the lowest RDI was considered as the least desirable in terms of AOA while the treatment with the highest RDI was considered as the most desirable one.

6.2.9. Statistical analysis

The data were submitted to ANOVA according to Kuehl (1999). Treatment effects were evaluated by F test at α=0.05. The results were reported as mean±SE. Means comparison was done by Duncan multiple range test at α level= 0.05. The relationship among the different variables analyzed in this study was measured by the Pearson correlation coefficient (R). The relative fit goodness of the different analytical methods used was measured by comparison of the coefficients of determination (R^2), and coefficients of variation (CV). Regression analyses were carried out to study the relationship of the total phenolic, total flavonoid, and total carotenoid contents with the AOA parameters. The SAS (Statistical Analysis System, v. 8.1; SAS Institute Inc., Cary, NC, USA) software was used.

6.3. Results

The effects of air-drying methods on color parameters, total phenolics, total flavonoids, total carotenoids, beta-carotene, and antioxidant activities of TNG57, TNG66, and TNG73 sweet potato were evaluated. Flesh samples of TNG57 and TNG66 were analyzed, while both flesh and skin samples of TNG73 were considered. The results are reported in this section.

6.3.1. Effect of the drying process on the color parameters

A) Cultivar TNG57

The yellow visible color of the fresh samples of TNG57 (h°=86.51) changed to blue and reddish yellow in freeze-drying and sun-drying, respectively, while it shifted to greenish yellow during low-temperature drying (25°C). During hot air drying, the color changed to bluish red at 50°C, and reddish blue at 75°C (Table 19). The color lightness of the fresh samples (L=71.62) decreased during sun-drying and low-temperature drying (25°C) and increased during freeze-drying and hot air drying. The color intensity which corresponded to a chroma value of 27.93 in the fresh samples decreased in all drying conditions. Figure 11 shows a visual color appearance of the samples of TNG57 before and after drying.

B) Cultivar TNG66

The reddish yellow color of TNG66 fresh samples (h°=56.73) shifted to yellow during freeze-drying, sun-drying, low-temperature drying (25°C), and hot hair drying at 50°C (Table 20). No significant change was observed in the visible color during the hot air drying at 75°C. The color lightness of the fresh samples (L=63.40) increased during freeze-drying and hot air drying at 50 and 75°C and decreased during sun-drying and low-temperature drying at 25°C. The color intensity decreased in all drying conditions. Figure 12 shows a visual color appearance of the samples before and after drying.

Table 19. Changes induced in hue angle, color lightness, and color intensity of sweet potato cultivar TNG57 by the drying process

Treatment[1]	(Mean±SE, n=3)[2]		
	Hue angle (h°)	Lightness (L)	Chroma(C)
Freeze-drying at -40°C	275.29± 0.03^{c}	87.22± 0.02^{a}	14.46± 0.07^{a}
Sun-drying	75.39±<0.01^{e}	55.34± 0.48^{c}	13.44±<0.00^{b}
LTD at 25°C	128.18±14.98^{d}	33.88±<0.01^{d}	6.66±<0.01^{c}
HAD at 50°C	332.84± 3.46^{a}	80.58±<0.01^{b}	15.21± 0.02^{a}
HAD at 75°C	313.19± 1.91^{b}	79.83± 0.01^{b}	14.84± 0.00^{a}

[1]LTD: low-temperature drying; HAD: hot air drying. [2]Means with a same letter in a column are not significantly different (p>0.05).

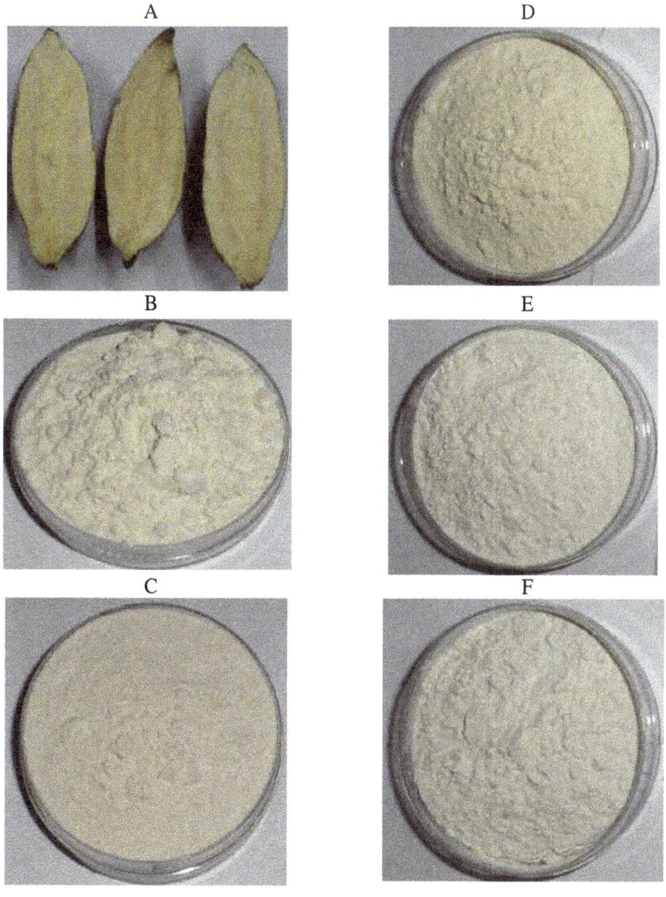

Figure 11. Visual color appearances of TNG57 before and after drying.

A: root cross section; B, C, D, E, F: powder of samples previously subjected to freeze-drying, sun-drying, low-temperature air-drying at 25°C, hot air drying (HAD) at 50°C, and HAD at 75°C, respectively.

Table 20. Changes induced in hue angle, color lightness, and color intensity of TNG66 by the drying process

Treatment[1]	(Mean±SE, n=3)[2]		
	Hue angle (h°)	Lightness (L)	Chroma(C)
Freeze-drying at -40°C	67.56±0.21[c]	74.65±1.70[a]	25.43±0.06[a]
Sun-drying	79.73±0.90[a]	56.57±0.01[b]	19.86±0.01[d]
LTD at 25°C	78.22±0.80[a]	25.29±0.00[c]	7.60±0.00[e]
HAD at 50°C	71.90±1.84[b]	74.80±0.00[a]	21.64±0.00[c]
HAD at 75°C	63.26±0.45[d]	75.90±0.02[a]	24.26±0.01[b]

[1]LTD: low-temperature drying; HAD: hot air drying. [2]Means with the same letter in a column are not significantly different ($p>0.05$).

Figure 12. Visual color appearances of TNG66 before and after drying.

A: root cross section; B, C, D, E, F: powder of samples previously subjected to freeze-drying, sun-drying, low-temperature air-drying at 25°C, hot air drying (HAD) at 50°C, and HAD at 75°C, respectively.

C) Cultivar TNG73

The data summarized in Table 21 show that the red color of the fresh flesh samples of the cultivar TNG73 (h°=354.54) only turned to reddish yellow during the drying at hot air at 50°C. The visible color was maintained during the drying process in the other conditions. The color lightness (L=25.36 in the fresh samples) increased during freeze-drying, sun-drying, and hot air drying at 50°C, while it decreased during drying at 25 and 75°C. The color intensity (C=34.98 in the fresh samples) decreased in all drying conditions. The skin color changed from yellowish red to red during freeze-drying and sun-drying. It changed to reddish yellow during LTD at 25°C and HAD both at 50 and 75°C. The color intensity and color lightness of the skin increased in all drying processes. Figure 13 shows the visual color appearance of the samples of TNG73 before and after drying.

Table 21. Changes induced in hue angle, color lightness, and color intensity of TNG73 by the drying process[1]

Treatment[1]	(Mean±SE, n=3)[2]		
	Hue angle (h°)	Lightness (L)	Chroma (C)
Flesh			
Freeze-drying at -40°C	344.26±0.04^{b}	51.32±0.05^{c}	28.25±0.09^{a}
Sun-drying	352.74±2.18^{a}	69.39±0.73^{a}	23.92±0.20^{b}
LTD at 25°C	352.16±0.07^{a}	21.42±0.00^{d}	6.74±0.01^{d}
HAD at 50°C	54.95±4.19^{c}	56.56±0.44^{b}	21.58±0.71^{c}
HAD at 75°C	9.67±0.11^{d}	23.37±0.00^{d}	3.81±0.01^{e}
Skin			
Freeze-drying at -40°C	349.12± 0.16^{a}	45.90±0.34^{e}	28.25±0.09^{a}
Sun-drying	328.31±10.13^{a}	64.93±0.78^{a}	22.37±0.06^{c}
LTD at 25°C	56.01± 3.50^{b}	48.99±0.29^{d}	26.96±0.23^{a}
HAD at 50°C	62.15± 4.94^{b}	56.12±0.26^{c}	22.61±0.71^{b}
HAD at 75°C	58.37± 6.97^{b}	59.57±0.85^{b}	28.40±0.90^{a}

[1]LTD: low-temperature drying; HAD: hot air drying. [2]Means with the same letter in a column are not significantly different (p>0.05).

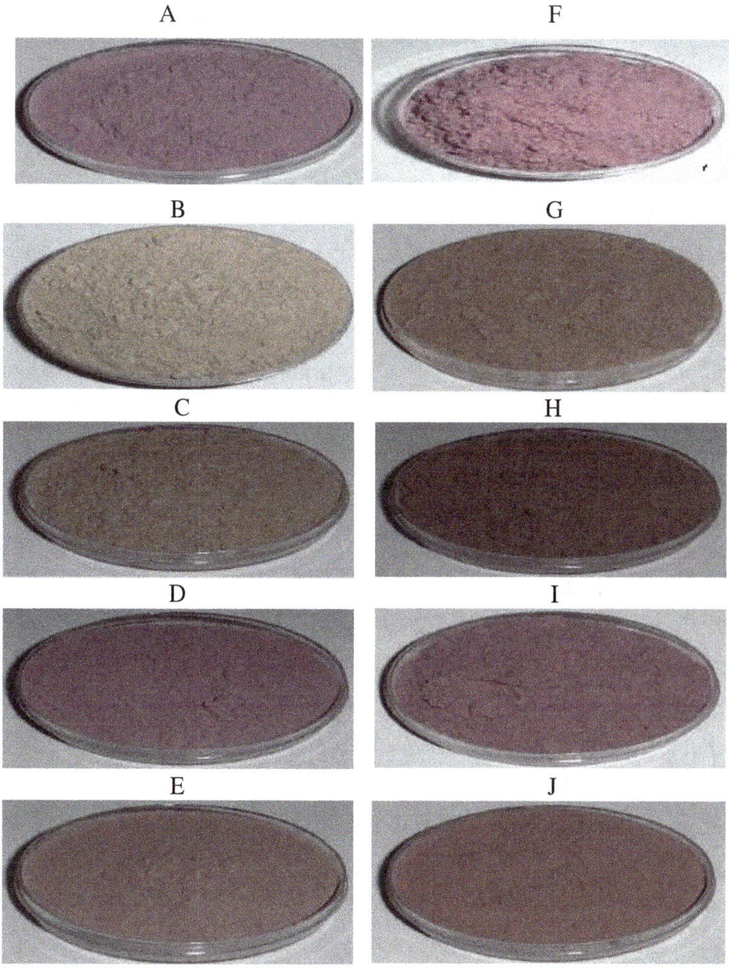

Figure 13. Visual color appearance of the TNG73 after drying at in different conditions.

A, B, C, D, E: powder of flesh subjected to freeze-drying (FZD), sun-drying, low-temperature air-drying (LTD) at 25°C, hot air drying (HAD) at 50°C, and HAD at 75°C, respectively. F, G, H, I, J: powder of skin subjected to FZD, sun-drying, LTD at 25°C, HAD at 50°C, and HAD at 75°C, respectively.

6.3.2. Effect of the drying process on total phenolic and flavonoid contents

A) Cultivar TNG57

The highest total phenolic content of TNG57 was observed in freeze-dried samples, followed by those dried at 25°C and in sunlight, then those dried at 25°C, and then those dried at 75°C (Table 22). All the air-drying conditions of TNG57 resulted in substantial losses of total phenolics from 61.03% in sun-drying to 82.89% in HAD at 75°C in comparison to the control, freeze-drying (Figure 14). Comparing to LTD at 25°C and sun-drying, HAD at 50 and 75°C resulted in higher losses of total phenolics. Similarly to total phenolics, the highest value of total flavonoids was observed in freeze-dried samples, followed by the sun-dried ones, then those dried at 25°C, and then those dried with hot air at 50 and 75°C (Table 25). The air drying process resulted in losses of total flavonoids from 59.97% in sun-drying to 81.48% in HAD at 75°C (Figure 14).

B) Cultivar TNG66

The total phenolic content of TNG66 decreased significantly in all drying conditions (Table 23). The highest value of total phenolics was observed in freeze-drying samples, followed by the sun-dried ones, then those dried at low-temperature at 25°C, then those dried with hot air at 75°C, and then those dried with hot air at 75°C. In LTD at 25°C and HAD at 50 and 75°C, the total phenolic content of the samples decreased with the increase of the drying temperature (Figure 15). The total flavonoid content showed changes similar to the total phenolic content during LTD at 25°C and HAD at 50 and 75°C. The highest flavonoids content was observed in freeze-dried samples, followed by those dried at low temperature at 25°C and those dried with hot air at 50°C, and then those dried with hot air at 75°C (Table 23). The highest decrease of total flavonoids was observed in sun-dried samples (60.66%), followed by those dried at 75°C (57.97%), then those dried at 50°C, and then those dried at 25°C (Figure 15).

Table 22. Evolution of the total phenolic and flavonoid contents of the flesh of TNG57 during the drying process

Treatment[1]	(Mean±SE, n=3)[2]	
	A	B
Freeze-drying at -40°C	48.21±1.56[a]	35.47±0.05[a]
Sun-drying	17.36±0.03[b]	14.20±0.04[b]
LTD at 25°C	18.79±0.01[b]	10.21±0.04[c]
HAD at 50°C	10.27±0.04[c]	6.90±0.03[d]
HAD at 75°C	8.25±0.11[d]	6.57±0.03[d]

[1]FZD: freeze-drying; LTD: low-temperature drying; HAD: hot air drying. [2]FZD: freeze-drying; LTD: low-temperature drying; HAD: hot air drying. [2]Means with a same letter in a column are not significantly different ($p>0.05$); A: total phenolics (mg gallic acid equivalent/100 g dry matter) B: total flavonoids (mg catechin equivalent/100 g sample dry matter).

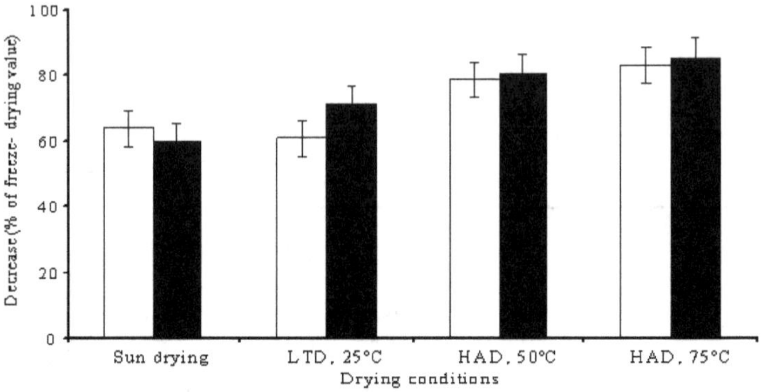

Figure 14. Decrease of the total phenolic and total flavonoid contents of TNG57 during air-drying with regards to the freeze-drying values. □: total phenolics, n: total flavonoids, LTD: low-temperature drying, HAD: hot air drying.

Table 23. Evolution of the total phenolic and total flavonoid contents of the flesh of TNG66 during the drying process

Treatment[1]	(Mean±SE, n=3)[2]	
	A	B
Freeze-drying at -40°C	160.56±1.95^{a}	105.75±0.23^{a}
Sun-drying	142.83±0.02^{b}	41.60±0.32^{d}
LTD at 25°C	128.84±0.05^{c}	76.26±0.15^{b}
HAD at 50°C	100.04±0.02^{d}	73.24±0.15^{b}
HAD at 75°C	50.85±0.06^{e}	44.45±0.11^{c}

[1]LTD: low-temperature drying; HAD: hot air drying. [2]Means with a same letter in a column are not significantly different (p>0.05); A: total phenolics (mg gallic acid equivalent/100 g sample dry matter) B: total flavonoids (mg catechin equivalent/100 g sample dry matter).

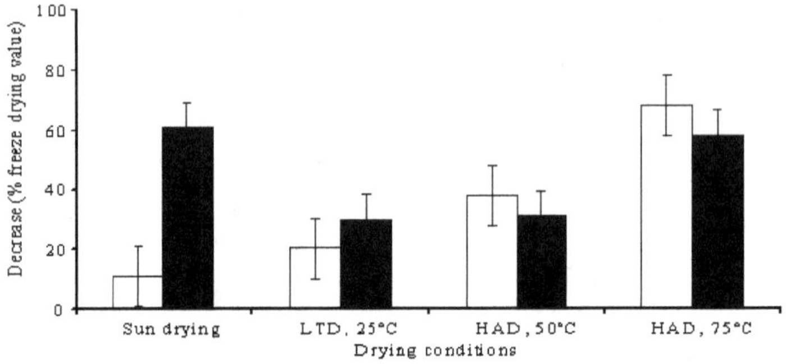

Figure 15. Decrease of the total phenolic and total flavonoid contents of TNG66 during air-drying with regards to the freeze-drying values.

□: total phenolics, n : total flavonoids, LTD: low-temperature drying, HAD: hot air drying.

C) Cultivar TNG73

Table 24 shows the total phenolic and total flavonoid contents of TNG73 sweet potato flesh and skin as affected by the drying method and conditions. In comparison to values obtained in freeze-drying used as control, the total phenolic content of TNG73 decreased significantly in all air drying conditions both for the flesh and the skin samples. Similarly to the total phenolic content, the total flavonoid content decreased significantly in all air drying conditions in comparison to freeze-drying. On average, drying TNG73 at low temperature for extended time resulted in higher losses of total phenolics and total flavonoids than drying at high temperature for short time. The skin consistently exhibited higher total phenolic and total flavonoid content than the flesh.

6.3.3. Effect of the drying process on the total carotenoid and beta-carotene contents

A) Cultivar TNG57

For TNG57, the freeze-dried samples had the highest carotenoid and beta-carotene contents, followed by the samples dried at 25°C, then those sun-dried, then those dried at 75°C, and then those dried at 50°C (Table 25). Beta-carotene content showed similar changes. The losses of total carotenoids and beta-carotene during the drying process were very high in all air-drying conditions.

B) Cultivar TNG66

TNG66 carotenoid content was highest in freeze-dried samples, followed by those dried at 25°C, then those sun-dried, and then those dried with hot air at 75°C (Table 26). The lowest values were observed in samples dried with hot air at 50°C. The beta-carotene content showed similar changes during the drying process. The HPLC chromatograms presented in Figure 16 show that beta-carotene was the main carotenoid of TNG66 sweet potato.

Table 24. Evolution of the total phenolic and flavonoid contents of the flesh of TNG73 during the drying process

Treatment[1]	(Mean±SE, n=3)[2]	
	A	B
Flesh		
Freeze-drying at -40°C	515.54±4.27^{a}	232.67±2.01^{a}
Sun-drying	24.81±0.11^{d}	20.97±0.07^{d}
LTD at 25°C	31.13±0.01^{c}	24.66±0.19^{c}
HAD at 50°C	82.66±0.02^{b}	37.42±0.02^{b}
HAD at 75°C	81.62±0.08^{b}	38.82±0.02^{b}
Skin		
Freeze-drying at -40°C	1758.22± 6.11^{a}	675.47±4.70^{a}
Sun-drying	79.28± 0.72^{e}	61.74±0.07^{e}
LTD at 25°C	102.50± 0.30^{d}	66.79±0.15^{d}
HAD at 50°C	201.14±26.47^{c}	76.68±0.13^{c}
HAD at 75°C	260.40± 1.22^{b}	109.44±0.02^{b}

[1]LTD: low-temperature drying; HAD: hot air drying. [2]Means with a same letter in a column are not significantly different (p>0.05); A: total phenolics (mg gallic acid equivalent/100 g sample dry matter); B: total flavonoids (mg catechin equivalent/100 g sample dry matter).

Table 25. Evolution of the total carotenoids and beta-carotene contents of the flesh of TNG57

Treatment[1]	(Mean±SE, n=3)[2]	
	A	B
Freeze-drying at -40°C	128.52±0.51^{a}	116.93±0.20^{a}
Sun-drying	2.80±0.02^{c}	2.67±0.01^{c}
LTD at 25°C	7.68±0.03^{b}	6.91±0.03^{b}
HAD at 50°C	1.67±0.00^{e}	1.56±0.04^{e}
HAD at 75°C	2.16±<0.01^{d}	1.93±0.03^{d}

[1]LTD: low-temperature drying; HAD: hot air drying. [2]Means with a same letter in a column are not significantly different (p>0.05); A: total carotenoids content (μg beta-carotene equivalent/g sample dry matter) and B: beta-carotene content (mg/kg sample dry matter).

Table 26. Total carotenoid and beta-carotene contents of TNG66 as affected by the drying process[1]

Treatment[1]	(Mean±SE, n=3)[2]	
	A	B
Freeze-drying at -40°C	504.42±1.09[a]	458.29±2.97[a]
Sun-drying	209.73±0.20[c]	200.37±0.71[c]
LTD at 25°C	223.63±0.29[b]	201.33± 0.87[b]
HAD at 50°C	48.36±0.64[e]	45.03±1.43[e]
HAD at 75°C	199.14±0.74[d]	177.54±4.43[d]

[1]LTD: low-temperature drying; HAD: hot air drying. [2]Means with a same letter in a column are not significantly different ($p>0.05$); A: total carotenoids content (μg beta-carotene equivalent/g sample dry matter) and B: beta-carotene content (mg/kg sample dry matter).

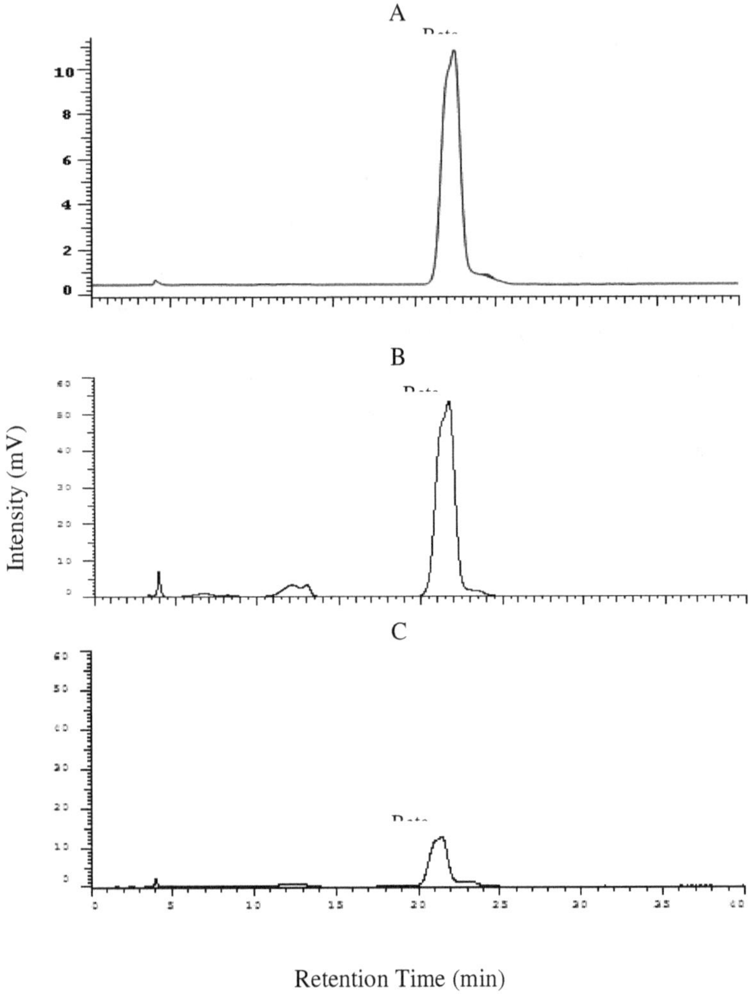

Figure 16. Representative chromatograms of a standard beta-carotene solution and fractions of TNG66.

A: standard; B, and C: fractions of TNG66 dried at 25°C and 50°C, respectively.

C) Cultivar TNG73

The results of carotenoid and beta-carotene content of TNG73 are summarized in Table 27. In all air-drying conditions, substantial losses of total carotenoids and beta-carotene were observed in comparison to the control (freeze-drying). For the flesh samples, the total carotenoid content varied from 2.29 in samples dried at 75°C to 30.51 μg beta-carotene equivalent /g dry matter in the control. For the skin samples, it varied from 2.91 μg beta-carotene equivalent /g dry matter in sun-drying to 17.17 μg beta-carotene equivalent /g dry matter in the control samples.

The beta-carotene content of the flesh samples ranged from 0.25 mg/kg dry matter in sun-drying to 2.08 mg/kg dry matter in the control. For the skin samples, beta-carotene content varied from 0.19 mg/kg dry matter in sun-drying to 1.15 mg/kg dry matter in the control. These results show that TNG73 is a low carotenoid content cultivar in comparison to results reported in the literature (Teow *et al.*, 2007; Duvivier *et al.,* 2008) and in comparison to TNG66.

Table 27. Total carotenoid and beta-carotene contents of TNG73 as affected by the drying process

Treatment[1]	(Mean±SE, n=3)[2]	
	Total carotenoids (mg beta-carotene equivalent/kg dry matter)	Beta-carotene (mg/kg dry matter)
Flesh		
Freeze-drying at -40°C	30.51±1.16[a]	2.08±0.10[a]
Sun-drying	2.58±0.01[d]	0.25±0.01[d]
LTD at 25°C	3.06±0.01[b]	1.63±0.07[b]
HAD at 50°C	2.68±0.00[c]	0.14±0.01[e]
HAD at 75°C	2.29±0.00[e]	0.35±.03[c]
Skin		
Freeze-drying at -40°C	17.17±0.03[a]	1.15±0.01[a]
Sun-drying	1.91±0.03[e]	0.19±0.00[d]
LTD at 25°C	3.50±0.03[d]	0.86±0.01[b]
HAD at 50°C	3.83±0.05[c]	0.20±0.02[d]
HAD at 75°C	4.16±0.04[b]	0.63±0.00[c]

[1]LTD: low-temperature drying; HAD: hot air drying. [2]Within samples, means with different letters in a column are significantly different ($p<0.05$).

6.3.4. Effect of the drying process on the antioxidant activity

A) Cultivar TNG57

The highest DPPH FRSA of TNG57 was observed in freeze-dried samples, followed by those dried at low temperature at 25°C, then those dried with hot air at 50°C, then those dried with hot air at 75°C, and then those sun-dried (Table 28). In drying at 25, 50, and 75°C, the DPPH FRSA decreased with the increasing drying temperature. Sun-drying and LTD resulted in decreases of the DPPH FRSA with regards to the control by 63.24 and 2.10 %, respectively, while in drying with hot air the decreases were 6.02 and 40.73 % at 50 and 75°C, respectively (Figure 17).

Freeze-drying and LTD at 25°C exhibited the highest AOP values, followed by HAD at 50 and 75°C, and then sun-drying (Table 28). The sun-drying and HAD at 50 and 75°C resulted in decreases of 62.63, 17.19, and 18.29 % of the AOP value, respectively, with regards to the control.

The highest ABTS FRSA of TNG57 was observed in freeze-dried samples, followed by those dried at low temperature at 25°C, then those dried with hot air at 50°C, then those dried with hot air at 75°C, and then those sun-dried (Table 28). HAD at 50 and 75°C resulted in losses of the ABTS FRSA of 82.06 and 88.03 %, respectively, with regards to the control, while the decreases in LTD at 25°C and sun-drying were respectively 78.81 and 92.85 %, respectively, with regards to the control (Figure 17).

Table 28. DPPH FRSA, AOP, and ABTS FRSA of TNG57 as affected by the drying process

Treatment[1]	(Mean±SE, n=3)[2]		
	DPPH FRSA (%)	AOP (%)	ABTS FRSA[3]
Freeze-drying at -40°C	90.52± 0.06^{a}	86.93±0.13^{a}	6013.98±371.21^{a}
Sun-drying	33.27±13.35^{e}	31.62±1.04^{c}	430.00±24.71^{e}
LTD at 25°C	88.62± 1.66^{b}	86.82±<0.00^{a}	1274.75±40.38^{b}
HAD at 50°C	77.03± 3.57^{c}	71.99±0.24^{b}	1066.93± 4.93^{c}
HAD at 75°C	48.58± 9.42^{d}	71.03±0.14^{b}	720.04±27.65^{d}

[1]LTD: low-temperature drying; HAD: hot air drying. [2]Means with a same letter in a column are not significantly different (p>0.05). [3]Results in mg vitamin C equivalent/100 g dry matter).

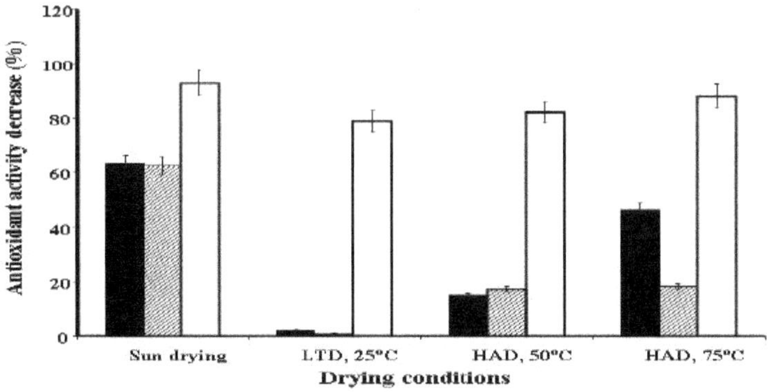

Figure 17. Decrease of the DPPH FRSA, AOP, and ABTS FRSA of TNG57 during air-drying with regards to the freeze-drying values.
n : DPPH FRSA, ▨: AOP, □: ABTS FRSA, LTD: low-temperature drying, HAD: hot air drying. The results are mean±SE of 3 replicates.

B) Cultivar TNG66

For the cultivar TNG66, the highest values of DPPH FRSA were observed in samples freeze-dried, those sun-dried, and those dried 25°C, followed by those dried with hot air at 50°C, and then those dried with hot air at 75°C (Table 29). Freeze-dried samples exhibited the highest values of AOP, followed in descending order by those treated with sun-drying, and then those submitted to LTD at 25°C, HAD at 50°C, and HAD at 75°C. The highest ABTS FRSA value was observed in freeze-dried samples, followed by those treated with sun-drying, while the samples subjected to HAD at 75°C showed the lowest ABTS FRSA. The samples submitted LTD at 25°C and HAD at 50°C were not significantly different in terms of ABTS FRSA.

In comparison to the values the control, freeze-drying, no significant decreases of the DPPH FRSA of the TNG66 samples treated with sun-drying and LTD at 25°C were observed. However, for TNG66 samples submitted to HAD at

50 and HAD at 75°C, the DPPH FRSA decreased by 19.02 and 45.17%, respectively (Figure 18). The decrease of AOP was low to moderate, amounting to 4.54, 17.07, 18.53, and 16.66 % in sun-drying, LTD at 25°C, HAD at 50°C, and HAD at 75°C, respectively. In contrast of the AOP decreases that did not reach 20%, substantial decreases of ABTS FRSA values were observed during the drying processes, ranging from 84.64 % in sun-drying to 95.58 % in HAD at 75°C.

Table 29. DPPH FRSA, AOP, and ABTS FRSA of TNG66 as affected by the drying process

Treatment[1]	(Mean±SE, n=3)[2]		
	DPPH FRSA (%)	AOP (%)	ABTS FRSA[3]
Freeze-drying at -40°C	91.61±0.33^{a}	65.67±0.86^{a}	1793.70±55.99^{a}
Sun-drying	91.86±0.19^{a}	62.69±3.43^{b}	278.62±2.54^{b}
LTD at 25°C	90.76±0.03^{a}	54.46±0.14^{c}	195.23±9.40^{c}
HAD at 50°C	74.19±0.23^{b}	53.50±0.24^{c}	209.72±1.81^{c}
HAD at 75°C	50.23±0.30^{c}	54.73±0.48^{c}	80.24±4.79^{d}

[1]LTD: low-temperature drying; HAD: hot air drying. [2]Means with the same small letter in a column are not significantly different (>0.05). [3]Results in mg vitamin C equivalent/100 g dry matter.

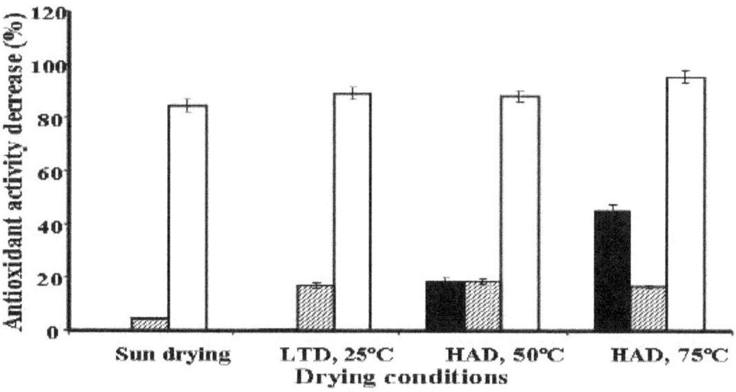

Figure 18. Decrease of the DPPH FRSA, AOP, and ABTS FRSA of TNG66 during air-drying with regards to the freeze-drying values. n : DPPH FRSA, ▨: AOP, □: ABTS FRSA, LTD: low-temperature drying HAD: hot air drying. The results are mean±SE of 3 replicates.

C) Cultivar TNG73

The effects of the drying process on the DPPH FRSA of TNG73 are summarized in Table 30. The highest DPPH FRSA of the flesh was observed in freeze dried samples, followed by those dried with hot air at 75°C, then those sun-dried, then those dried hot air-dried at 50°C, and then those dried at 25°C. Freeze-dried flesh samples of TNG73 exhibited the highest AOP values, followed by those dried with hot air at 50 and 75°C, and then those submitted to sun-drying and LTD at 25°C. The highest ABTS FRSA value for the flesh was observed in the freeze-dried samples, followed by those treated with HAD at 75°C, then those submitted to sun-drying and those subjected to LTD at 25°C, and then those treated with HAD at 50°C. In contrast of the flesh samples in which the DPPH FRSA varied from 46.79% in LTD at 25°C to 90.32% in in freeze-drying, the skin samples showed DPPH FRSA higher than 80% all drying treatments. The DPPH FRSA of the skin samples submitted to freeze-drying, sun-drying, and LTD at 25°C exceeded 90%.

The highest DPPH FRSA of the skin was observed in sun-dried samples, followed by those treated with freeze-dying and those subjected to LTD at 25°C, then those submitted to HAD at 75°C and those submitted to HAD at 50°C. Freeze-dried and sun-dried skin samples of TNG73 exhibited the highest AOP values, followed by those treated with HAD at 50°C, and those treated with HAD at 75°C, and then those submitted to sun-drying and those submitted to LTD at 25°C. The highest ABTS FRSA value for the skin was observed in the freeze-dried samples, followed by those treated with HAD at 75°C, then those subjected to LTD at 25°C, then those submitted to sun-drying, and then those treated with HAD at 50°C.

Table 30. DPPH FRSA, AOP, and ABTS FRSA of TNG73 as affected by the drying process

Treatment[1]	(Mean±SE, n=3)[2]		
	DPPH FRSA (%)	AOP (%)	ABTS FRSA[3]
Flesh			
Freeze-drying, -40°C	92.23±0.25[a]	73.00±1.15[a]	4552.67±28.35[a]
Sun-drying	86.74±0.08[b]	64.07±0.00[c]	616.52±0.41[c]
LTD at 25°C	46.79±0.39[d]	59.18±0.71[c]	616.04±1.75[c]
HAD at 50°C	77.33±0.00[c]	68.46±0.20[b]	569.68±0.00[d]
HAD at 75°C	90.32±0.00[a]	67.66±0.00[b]	1524.60±1.82[b]
Skin			
Freeze-drying, -40°C	91.17±0.10[b]	80.48±1.25[a]	4355.94±23.31[a]
Sun-drying	94.06±0.08[a]	80.44±0.20[a]	1618.02±4.56[d]
LTD at 25°C	90.59±0.04[b]	62.87±1.20[c]	1898.97±0.88[c]
HAD at 50°C	81.11±0.12[d]	73.65±0.92[b]	622.04±1.70[e]
HAD at 75°C	89.10±0.16[c]	74.85±0.60[b]	3224.73±0.00[b]

[1]LTD: low-temperature drying; HAD: hot air drying. [2]Within samples, means with different letters in a column are significantly different ($p<0.05$); [3]results in mg vitamin C equivalent /100 g dry matter.

Comparing to the values of the control, freeze-drying, the losses of DPPH FRSA in flesh samples submitted to sun-drying and LTD at 25°C were 5.95 and 49.92 %, respectively, while HAD at 50°C and HAD at 75°C resulted in losses of 16.16 and 2.07 %, respectively (Figure 19). The losses of AOP in flesh samples submitted to sun-drying, LTD at 25°C, HAD at 50°C, and HAD at 75°C were

12.23, 18.93, 6.22, and 7.33 %, respectively. All the air-drying processes of the flesh samples of TNG73 resulted in substantial decreases of ABTSA FRSA, from 66.51 % in HAD at 75°C, to 87.49 % in HAD at 50°C.

For the skin samples, Figure 19 shows that the highest decrease of the DPPH FRSA was observed in HAD at 50ºC (11.03%), followed in descending order by HAD at 75ºC (2.27%), then LTD at 25ºC (0.64%). No decrease of DPPH FRSA was observed during sun-drying of the skin samples of TNG73. The decrease of the AOP of the skin samples varied from 0.05% in sun drying to 21.88% in LTD at 25ºC. During HAD at 50ºC and HAD at 75ºC, the AOP of the skin samples decreased by 8.49 and 7.00%, respectively. All the air-drying processes of the skin samples of TNG73 resulted in significant decreases of ABTSA FRSA, from 25.97 % in HAD at 75°C, to 85.72 % in HAD at 50°C.

The data discussed in this section, and summarized in Tables 28 to 30 and Figures 17 to 19 show large variations in the losses of AOA in sweet potato during drying. The extent of AOA losses varied among TNG57, TNG66, and TNG73 cultivars and with the methods used to assess the AOA. As the three cultivars showed different contents of phenolics and carotenoids, the correlations of the AOA with the contents of total phenolics, total flavonoids, total carotenoids, and β-carotene was further evaluated. The correlations among the three methods used to assess the AOA were also evaluated.

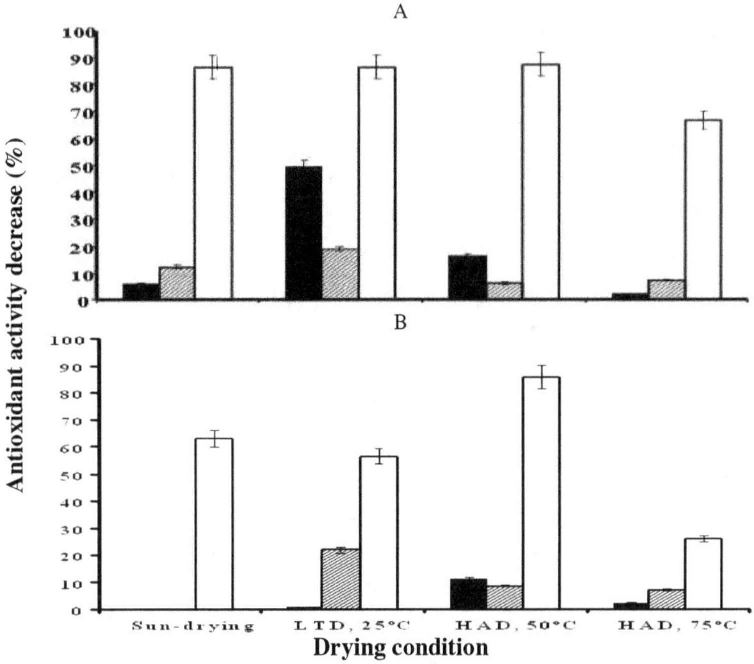

Figure 19. Decrease of the DPPH FRSA, AOP, and ABTS FRSA of TNG73 during air-drying with regards to the freeze-drying values.

A: flesh samples; B: skin samples; n : DPPH FRSA, ▨: AOP, □: ABTS FRSA, LTD: low-temperature drying, HAD: hot air drying. The results are mean±SE of 3 replicates.

6.3.5. Correlations among the antioxidant activity indicators, total phenolic, flavonoid, carotenoid, and β-carotene contents

A) Cultivar TNG57

Significant and positive correlations of total phenolics, total flavonoids, and total carotenoids were observed with ABTS FRSA (Table 31). However, the pair-wise correlations among the analytical methods (DPPH, AOP, and ABTS) as well as the correlations of total phenolics, total flavonoids, and total carotenoids with DPPH FRSA and AOP were not significant at 5% of probability. In fact, each assay involves different chemical reactions. A food crop that shows high AOA when analyzed with a method can be considered as a good candidate for functional food. However, the reciprocal is not true, that is, a food crop that does not show high AOA when analyzed by a specific method cannot be considered as unable to be used as functional food. Results from different methods can show large discrepancies. That is the reason for the lack of correlation among analytical methods poses a problem in ranking foods on basis of AOA.

B) Cultivar TNG66

The AOA measured as ABTS FRSA correlated positively ($R>0$) and significantly ($p<0.05$) with total phenolics (Table 32). However, the correlations with with total flavonoids and total carotenoids were not significant. The Pearson correlation coefficients listed in Table 36 show no significant correlations among the three analytical methods of evaluation of AOA.

Table 31. Matrix of coefficients of correlation (R) among different variables analyzed for TNG57 samples

Variables[1]	A	B	C	D	E	F
B	0.998**					
C	0.996**	0.989**				
D	0.322 NS	0.271 NS	0.369 NS			
E	0.998**	0.997**	0.995**	0.301 NS		
F	0.606 NS	0.616 NS	0.619 NS	0.239 NS	0.608 NS	
G	0.745 NS	0.705 NS	0.790*	0.833*	0.732 NS	0.305 NS

[1]A: total phenolics; B: total flavonoids; C: total carotenoids; D: DPPH FRSA; E: ABTS FRSA; F: AOP; G: Relative desirability index (RDI). *significant correlation ($p<0.05$); **highly significant correlation ($p<0.01$); NS non-significant correlation ($p>0.05$).

Table 32. Matrix of coefficients of correlation (R) among different variables analyzed for TNG66 samples

Variable[1]	A	B	C	D	E	F
B	0.989*					
C	0.829 NS	0.875 NS				
D	0.596 NS	0.501 NS	0.047 NS			
E	0.976*	0.932 NS	0.743 NS	0.690 NS		
F	0.814 NS	0.729 NS	0.618 NS	0.611 NS	0.912 NS	
G	0.875 NS	0.794 NS	0.509 NS	0.854 NS	0.953*	0.918 NS

[1]A: total phenolics; B: total flavonoids; C: total carotenoids; D: DPPH FRSA; E: ABTS FRSA; F: AOP; G: Relative desirability index (RDI). *significant correlation ($p<0.05$); **highly significant correlation ($p<0.01$); NS non-significant correlation ($p>0.05$).

C) Cultivar TNG73

The correlation of the AOA with the total phenolics was positive and significant in the three assays, while the correlation of the AOA with total carotenoids was significant only in ABTS assay (Table 33). The Pearson correlation coefficients listed in Table 33 show that the correlations among the three methods were positive (R>0) and significant. However, statistical significance of a correlation does not mean a high correlation. In fact, even statistically significant, the correlations among DPPH, ABTS, and AOP assays were low with the highest coefficient of correlation (between DPPH and AOP assays) equal to 0.61.

For TNG57 and TNG 66, no significant correlations were observed amnong the three assays used to evaluate the AOA, namely, DPPH, ABTS, and AOP assays. The correlations were significant in TNG73, but the coefficient of correlation were low (<0.70). Due to the lack of correlation among the three methods, we further evaluated their relative fit goodness on basis of the coefficient of determination (R^2) and coefficient of variation (CV).

6.3.6. Relative fit goodness of the three analytical methods

A) Cultivar TNG57

The regression parameters corresponding to the relative fit goodness of the different methods [coefficient of determination (R^2), coefficient of variation (CV), and the p-value] used to compute the AOA of TNG57 sweet potato are summarized in Table 34. AOP assay fitted better to study the AOA of TNG57 samples (R^2=0.9990; CV=11.90%; and p-value=0.0000). DPPH assay fitted less (R^2=0.834; CV=18.99%; p-value=0.0007). With R^2=0.859, CV=16.4%, and p-value=0.0003, ABTS assay showed intermediate fit goodness. The tests were highly significant with p<0.01 in all cases. However, the high values of R^2 (≥0.8340) and low values of CV (≤18.90%) indicated that the three methods are reliable to study the AOA of TNG57 sweet potato, even though they poorly correlated.

Table 33. Matrix of coefficients of correlation (R) among different variables analyzed for TNG73 samples

Variables[1]	A	B	C	D	E	F
B	0.64**					
C	0.75**	0.16^{NS}				
D	0.40*	0.36^{NS}	0.21^{NS}			
E	0.78**	0.31^{NS}	0.80**	0.44*		
F	0.63**	0.52**	0.23^{NS}	0.61**	0.37*	
G	0.45*	0.22^{NS}	0.38*	0.60*	0.63*	0.55*

[1]A: total phenolics; B: total flavonoids; C: total carotenoids; D: DPPH FRSA; E: ABTS FRSA; F: AOP; G: Relative desirability index (RDI). *significant correlation ($p<0.05$); **highly significant correlation ($p<0.01$); ^{NS} non-significant correlation ($p>0.05$).

Table 34. Regression parameters for relative fit goodness of DPPH, AOP, and ABTS assays in TNG57

Regression parameters	Assay		
	DPPH	AOP	ABTS
R^2 (accuracy)	0.8340	0.9990	0.8590
CV (%, precision)	18.9900	11.9000	16.4200
P-value (test significance)	0.0007	0.0000	0.0003

B) Cultivar TNG66

The regression parameters corresponding to the relative fit goodness of the different methods [coefficient of determination (R^2), coefficient of variation (CV), and the p-value] used to compute the AOA of TNG66 sweet potato are summarized in Table 35. For TNG66, DPPH assay fitted better with a coefficient of determination (R^2) =0.9990, a coefficient of variation (CV)=0.61%, and a p-value=0.0001. It was followed by ABTS assay (R^2=0.9920, CV=8.37%, and p-value=0.0001). AOP assay showed poorer fit goodness with values of R^2, CV, and p equal 0.8580, 4.62%, and 0.0030 respectively. However, as the lowest value of R^2 was 0.858 and the highest value of CV was 8.37%, we can consider that the three assays fitted to evaluate the AOA of TNG66 sweet potato.

C) Cultivar TNG73

The regression parameters corresponding to the relative fit goodness of the different methods [coefficient of determination (R^2), coefficient of variation (CV), and the p-value] used to compute the AOA of TNG73 sweet potato are summarized in Table 36. The three methods fitted well to study the AOA of sweet potato with R^2=0.9997, 0.9999, and 0.9967 for DPPH, ABTS, and AOP assays, respectively. These parameter's values mean little experiment error and high accuracy. The precision was also high in the three cases (CV=0.31, 2.12, and 0.35% in DPPH, AOP, and ABTS assays, respectively). The tests were highly significant with $p<0.01$ in all cases. Thus, we can conclude that the three methods are reliable to study the AOA of TNG73 sweet potato, even though they poorly correlated. Therefore, there is no sufficient evidence, to conclude that one the the three assays is the most or the least appropriate to study the AOA of sweet potato. Hence, we determined a single indicator of the AOA, relative desirability index (RDI), combining the the results of the three assays (DPPH FRSA, AOP, and ABTS FRSA). Furthermore, we evaluated the effect of the different drying methods and conditions on the RDI, for final ranking of the drying methods and conditions.

Table 35. Regression parameters for relative fit goodness of DPPH, AOP, and ABTS assays in TNG66

Regression parameters	Assay		
	DPPH	AOP	ABTS
R^2 (accuracy)	0.9990	0.8580	0.9920
CV (%, precision)	0.6090	4.6180	8.3700
P-value (test significance)	0.0001	0.0030	0.0001

Table 36. Regression parameters for relative fit goodness of DPPH, AOP, and ABTS assays in TNG73

Regression parameters	Assay		
	DPPH	AOP	ABTS
R^2 (accuracy)	0.9999	0.9967	0.9997
CV (%, precision)	0.3500	2.1200	0.3100
P-value (test significance)	0.0000	0.0000	0.0000

6.3.7. Effect of the drying process on the RDI Value

A) Cultivar TNG57

According to the RDI values, the freeze-dried samples of TNG57, were the most desirable, followed by the samples dried at 25°C, then those dried at 50°C, and then those dried at 75°C. Sun-dried samples were the least desirable (Figure 20). The RDI data correlated positively and significantly with those of DPPH assay (Table 31). Significant and positive pair-wise correlations of the RDI with total carotenoids of TNG57 samples were also observed. The correlations of RDI with total phenolics and total flavonoids were lower.

B) Cultivar TNG66

For TNG66, the RDI values indicated that the freeze-dried samples were the most desirable, followed by the sun-dried ones. In LTD at 25°C, HAD at 50°C, and HAD at 75°C, the RDI decreased with increasing drying temperature (Figure 21). The RDI data correlated positively and significantly with those of ABTS assay (Table 32).

C) Cultivar TNG73

Freeze-dried samples were the most desirable (Figure 22). For the flesh samples subjected to LTD at 25°C, HAD at 50°C, and HAD at 75°C, higher drying temperature combined with shorter drying time resulted in higher AOA in comparison to lower temperature combined to more extended drying time. The lowest RDI values were observed in flesh samples submitted to LTD at 25°C. The skin samples submitted to sun-drying, LTD at 25°C, HAD at 50°C, and HAD at 75°C showed similar RDI values. The RDI data correlated positively and significantly with those of DPPH, ABTS, and AOP assays. The RDI values also correlated positively and significantly with the total phenolics and the total carotenoids, the correlation with the total phenolics being higher (Table 33).

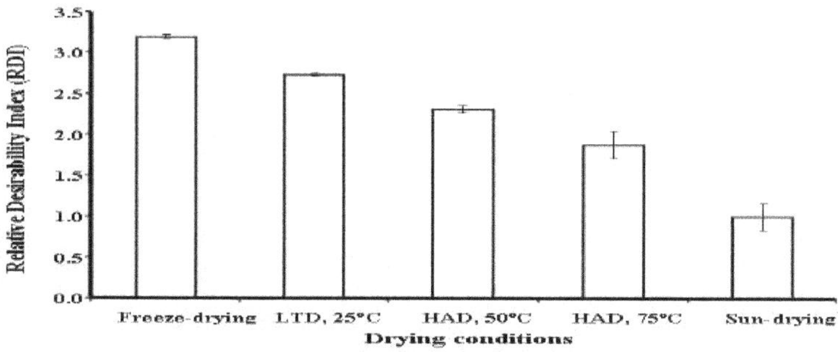

Figure 20. Antioxidant activity relative desirability index (RDI) of the flesh of TNG57 submitted to different drying processes.
The results are mean±SE of 3 replicates.

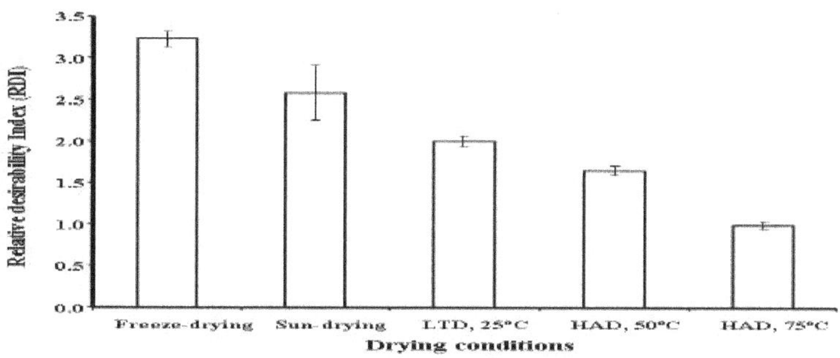

Figure 21. Antioxidant activity relative desirability index (RDI) of the flesh of TNG66 submitted to different drying processes.
The results are mean±SE of 3 replicates.

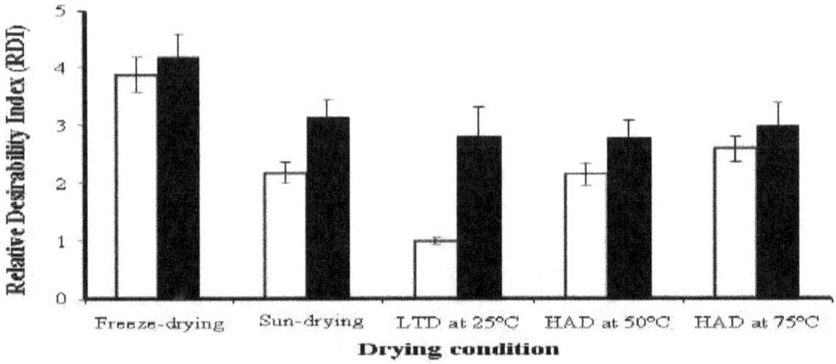

Figure 22. Antioxidant activity relative desirability index (RDI) of TNG73 submitted to different drying processes.
□: flesh, ■: skin; the results are mean±SE of 3 replicates.

6.4. Discussion

All the drying processes resulted in changes of the original color of the samples. The color intensity (chroma value) decreased, and the hue angle increased in all drying conditions of the three cultivars. The lightness increased in freeze-drying and hot air-drying, while it decreased in sun-drying and LTD of TNG57 and TNG66. This is probably due to longer exposure to oxygen. For TNG73, the lightness decreased in freeze-drying of the skin samples and LTD, and increased in freeze-drying of the flesh samples and HAD at 50°C, remaining stable in sun-drying and HAD at 75°C. These results suggested that the changes of the color during the drying process are genotype-dependent.

The losses of color intensity (yellowness in TNG57 and TNG66, and redness in TNG73) were accompanied with losses AOA. In Fact, the RDI decreased in all air-drying processes, showing losses of AOA with regards to the freeze-drying, in which moderate losses of AOA was observed with regards to the raw samples. Those results agreed with Rehman *et al.* (2003) who reported that evaporation and decomposition at elevated temperatures results in loss of AOA. Similarly, Larrauri *et al.* (1998) observed reduction of AOA by 28% in the red grape pomace peels when dried with hot air at 100°C.

The variation of the phenolic content during the drying processes paralleled with that of the AOA. The significant pair-wise correlations observed between total phenolics and the AOA of the three cultivars agreed with Islam *et al.* (2003) who reported significant correlation between the AOA and the total phenolics content of sweet potato leaves. The parameters of pair-wise correlations involving total phenolics, total flavonoids, and total carotenoids indicated that the drying process affected these different secondary metabolites in similar way.

6.5. Conclusions

The color intensity (chroma value) decreased, and the hue angle increased in all drying conditions of the three cultivars. Significant losses of color intensity (yellowness in TNG57, and redness in TNG66 and TNG73) were observed during the drying processes and accompanied with losses AOA, phenolics and carotenoid content. In general, the pair-wise correlations among DPPH, AOP, and ABTS assays were not high, and a single AOA single indicator, RDI, was computed to facilitate the ranking of the treatments. The effects of the drying process on the AOA were genotype-dependant. TNG73 showed higher suitability for drying at high temperature for short time than TNG57 and TNG66 in terms of retention of AOA. Higher correlations of the AOA were observed with the phenolic than carotenoid contents of TNG66 and TNG73, while the AOA of TNG57 correlated better with its carotenoid than its phenolic content.

Chapter 7. Extraction, HPLC separation, and characterization of the major phenolic antioxidants of TNG73 sweet potato (*Ipomoea batatas* Lam.) cultivar

Abstract

The phenolic profile of sweet potato (*Ipomoea batatas* Lam.) cv TNG73 and the contribution of each major individual phenolic compound in the profile to the overall antioxidant activity (AOA) of TNG73 were investigated. An ethanol crude extract of TNG73 was fractionated by liquid-liquid partition (LLP) using diethyl ether (DE) and ethyl acetate (EA) and by pH modification (PHM) to separate the crude extract into three fractions: acid-hydrolyzable phenolic acid fraction (AHPA), base-hydrolyzable phenolic acid fraction (BHPA), and free phenolic acid fraction (FPA). The fractions were analyzed by HPLC and the AOA was assessed by the 1, 1-diphenyl-2-picrylhydradzyl (DPPH) assay. Catechin, rutin, and gallic acid were the main phenolic compounds identified. Catechin was at the highest concentration in the BHPA fraction, whereas gallic acid was the major phenolic compound found in the acid-treated fraction. Rutin was found in free form in the DE fraction and was the major contributor to the AOA, followed by catechin and gallic acid.

Key words: antioxidant activity, catechin, rutin, gallic acid, sweet potato

7.1. Introduction

Sweet potato (*Ipomoea batatas* Lam.) is an easy-to-grow crop with good adaptability to diverse environmental conditions, high yielding ability and high energy content (Ravindran *et al.*, 1995). It is suitable for organic food production and other environmentally friendly agricultural practices. According to International Center of Potato (CIP, 1998), sweet potato ranks fifth among the most important food crops, in terms of fresh matter, after rice, wheat, corn, and cassava in developing countries where 95% of more than 133 million tons was annually produced. As such, sweet protato is considered as a promising crop for global food security (Scott, 1992). In addition to their nutritional values, some varieties are known to possess anti-carcinogenic properties and/or antioxidant activity (AOA) (Teow *et al.*, 2007). TNG73 is one of the most important sweet potato cultivars developed and grown in Taiwan besides TNG57, TNG66, Tainong 68, Tainong 70, and Taoyuan No.1. TNG73 has red color flesh (Figure 15) that makes it very attractive to consumers.

Previous experiments showed that it has high AOA, as measured by 1, 1-diphenyl-2-picrylhydradzyl (DPPH); anti-oxidative potency in linoleic acid system model (AOP), and 2, 2'-Azino-di-[3-ethylbenzothiazoline-6-sulfonate] (ABTS) assays. Its DPPH inhibition exceeded 90% in ethanol extract, and it maintained high AOA after drying at 75°C for 24 h. The AOA of TNG73 was attributed to high content of total phenolics, reaching 318.79 mg gallic acid equivalent /100 g sample fresh matter when extracted with ETOH (Duvivier *et al.*, 2008a). Its phenolic content is comparable to that of strawberries (Asami *et al.*, 2003). According to Kurata *et al.* (2006), sweet potato leaves have a high content of phenolics responsible for its anti-carcinogenic properties. Teow *et al.* (2007) also reported significant correlation between the AOA of tuberous root of sweet potato and its total phenolic content.

Phenolic compounds are known to comprise of more than 8000 biologically active substances, ranging from simple phenol molecule to polymeric structures with molecular mass above 30000 Da (Marinova *et al.*, 2005). On basis of the number of simple phenol units in the chemical structures, phenolic compounds are classified in two basic groups: simple phenols and poly-phenols that contain at least two phenol rings. The group of simple phenols contains also the so-called phenolic acids with carboxylic groups, while flavonoids are the main group of poly-phenols. Phenolic acids can be classified in benzoic acids and their derivatives and cinnamic acids and their derivatives. The main subclasses of flavonoids are flavanols (catechins and related compounds; and quercetins and their related glycosides), flavones (apigenin and luteolin), flavanonas (naringin) and isoflavones (genistein and daidzein). All of them occur mostly in nature as glycoside structures (Blasco *et al.*, 2005). More than 5000 flavonoid structures are currently known in higher and lower plants, indicating the chemical complexity of these compounds.

Determination of the antioxidant properties of TNG73 can elucidate its potential not only for development into a functional food, but also for use as an antioxidant source alternative to that of synthetic, which is commonly found in food industry. Thus, we analyzed the phenolic profile of TNG73 and evaluated the contribution of each major individual phenolic compound to its AOA.

7.2. Materials and Methods

7.2.1. Materials

Fresh sweet potato tuberous roots were purchased from a local market in Pingtung, Taiwan, in the June 2008. The samples were processed and analyzed immediately without being subjected to storage after they were brought back to the laboratory from the market. DE, hydrogen chloride (HCl), DPPH radical, catechin, gallic acid, rutin, sodium hydroxide (NaOH), and vitamin C were purchased from Sigma Chemical Co. (St. Louis, MO, USA). All the other reagents were either extra pure or analytical grade.

7.2.2. Sample preparation

Samples were washed in tap water, dried at room temperature, peeled, and cut in small pieces (0.5 to 1 cm of thickness). The sample dry mater (SDM) was measured according to the methods of the Association of Official Analytical Chemists (AOAC, 1970). Water activity was measured in an Aw Quick water activity meter (Rotronic Instrument Corp., Huntington, N.Y., U.S.A.). Flesh and skin samples were then dried at 75°C for one day. The dried samples were ground into a fine powder using an electric laboratory mill. The powder was weighted and sealed in laminated plastic bag and kept in desiccators until extraction.

7.2.3. Extraction

An aliquot of 10 ±0.01 g sample dry matter was mixed with 40 ml hexane, shaken for 1 h at room temperature at 100 rpm, and filtered through # 1 filter paper (Whatman Inc., Clifton, NJ). The procedure was repeated one more time for complete defatting. The defatted residue was dried under hood at room temperature over night and mixed with 100 ml ETOH in a 500 ml flask. The flask was then shaken at 100 rpm in darkness for 24 h in a water bath set at 25°C for complete extraction (Huang *et al.*, 2004). The mixture was filtered through Whatman no. 1 filter. The residue was washed with an additional 100 ml ETOH. The filtrate was evaporated to dryness using a rotary evaporator model BUCHI 111 equipped with a water-bath BUCHI 461 (Switzerland) and a water vacuum pump EYELA A-3S (Tokyo, Japan) at 40°C. The dried residue, referred to as crude extract, was weighed, re-dissolved in 10 ml 50% aqueous ETOH, and kept at -20°C.

7.2.4. Fractionation

The ETOH extract was fractionated using liquid-liquid partition (LLP) and pH modification (PHM) by acid-base treatment. The LLP method was adapted from the procedure described by William and Douglas (2006). An aliquot of 1 ml extract was diluted to 4 ml with 50% aqueous ETOH, mixed in a centrifuge tube with an equal volume of DE, vigorously shaken, and left to stand until phase separation. The DE layer was transferred into another tube using a dropper. The procedure was repeated two more times. The aqueous ETOH layer was then fractionated three times with EA. Figure 23 illustrates the fractionation process by LLP.

PHM was done according to Kim *et al.* (2005). An aliquot of 1 ml ETOH extract was evaporated to dryness using a rotary evaporator at 40°C. The residue was suspended in 4 ml acidified water (pH 2) with HCl and fractionated with 4 ml DE three times. This fraction contained free phenolic compounds (FPA). The aqueous layer was adjusted to pH 7 with 2 M NaOH and dried using a rotary evaporator at 40°C. The residue was re-dissolved in 4 ml 2 M NaOH, stirred for 4 h at room temperature, acidified to pH 2 with HCl, and fractionated with 4 ml DE three times. This fraction contained base-hydrolysable phenolic compounds (BHPA). The remaining aqueous layer was mixed with 3 ml 6 M HCl, heated at 95°C for 20 min, and fractionated with 4 ml DE three times. This fraction contained acid-hydrolysable phenolic compounds (AHPA). Figure 24 illustrates the fractionation process by PHM.

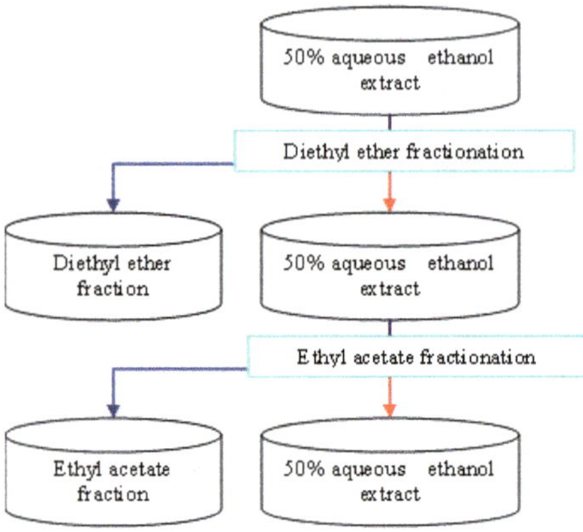

Figure 23. Flowchart of fractionation of TNG7373 ethanol extract by liquid-liquid partition using immiscible solvents

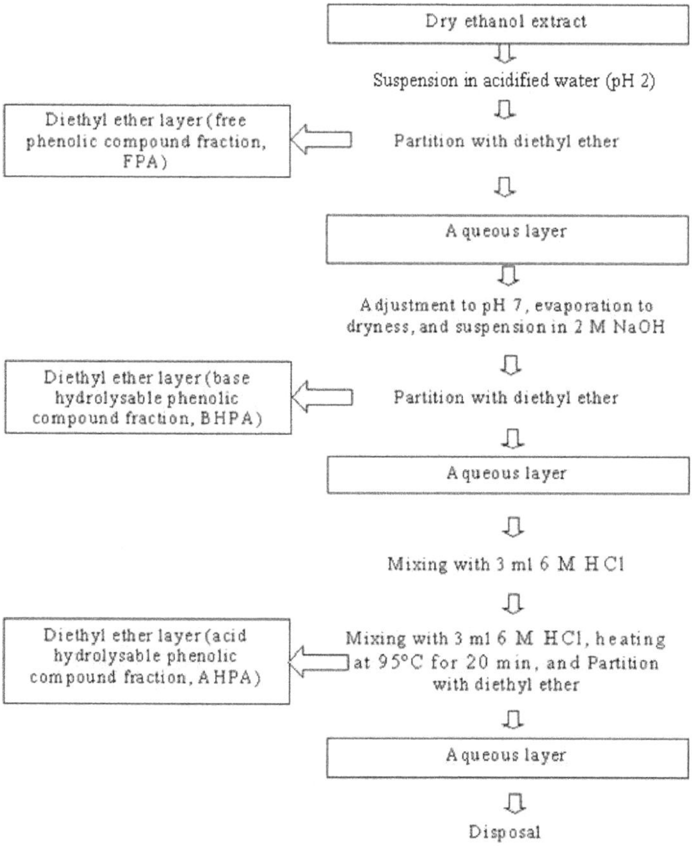

Figure 24. Flowchart of fractionation of TNG73 ethanol extract by pH modification

Each fraction (AHPA, BHPA, DE, EA, or FPA) was evaporated to dryness using a rotary evaporator at 40°C, re-dissolved in 10 ml MEOH, filtered through a 0.45 μm Millipore membrane, and kept at -20ºC until analysis. An aliquot of 1 ml was dried in a previously dried and weighed aluminum dish and the weight of the dried residue was recorded as total dry matter in 1 ml fraction.

7.2.5. Experiment Design

The study was realized in two steps. In the first step, five fractions (AHPA, BHPA, DE, EA, FPA) of the skin and the flesh of sweet potato cv TNG73 were analyzed for phenolic profile and individual phenolics' contents. A 2 * 5 factorial experiment was carried out in a completely randomized design (CRD) with two replicates, yielding 10 treatments and 20 experiment units. In this arrangement, the first factor refers to two parts of the root (flesh and skin), and the second refers to the five fractions. In the second step, the AOA of seven concentrations (0.05, 0.5, 5, 50, 125, 250, and 500 mg/l) of the five fractions of the skin and the flesh was studied. A 7*5*2 factorial experiment was used in a CRD with three replicates, yielding 70 treatments and 210 experiment units. In this arrangement, the first factor refers to the seven extract concentrations, the second refers to the two parts of the root, and the third refers to the five fractions.

7.2.6. HPLC separation and phenolic profile analysis

The phenolic profile was analyzed by HPLC using the procedure described in Liu *et al.* (2008). An aliquot of 20 μL sample previously filtered through a 0.45 μm Millipore membrane was injected in two replications into the liquid chromatographic system consisting of a high-pressure pump (HITACHI L-7100, Japan) coupled with a UV-visible detector (HITACHI L-7420, Japan). Separation was carried out in a RP18 column (Microsorb, 5 μm, 4 nm x 250 nm). Wavelength detection was 254 nm and the mobile phase system consisted of water (mobile phase A) and 5% acetic acid in methanol (mobile phase B). A gradient elution mode with decreasing polarity was used. The elution gradient was as follows: 20% B at 10 min, 40% B at 20 min, 50% B at 30 min, 70% B at 40 min, 80% B at 50 min, and 100% B at 60 min. The system was left to be reconditioned with 100% A for 15 min (0% B at 65 min, 0% B at 80%). The flow rate was 1.0 ml/min.

Identification was done by comparing peak retention times of unknown components and those of standards.

Quantification was done using standard calibration curves with the peak area as independent variable and concentration as dependant variable. The results were obtained in mg/l, and then converted into mg/kg dry matter when appropriate.

7.2.7. Antioxidant activity (AOA) assay

For each fraction, the AOA was measured by DPPH assay. DPPH assay was conducted according to Molyneux (2004). Absorbencies were monitored at 517 nm (A_{517}) in a spectrophotometer Beckman Coulter DU 730 UV/VIS and DPPH FRSA was determined using the following formula: DPPH FRSA (%) = $[(A_0-A_f)/A_0]*100$, where A_0 is the blank absorbance and A_f, the test sample absorbance. The DPPH FRSA values were plotted against the concentrations for each fraction. The DPPH IC50 value, that is, the concentration (mg/l) that produced 50% DPPH inhibition, was computed by graphic interpolation according to Stock (1939). The DPPH IC_{50} ratio of the crude extract to each fraction was computed to evaluate the enhancement of the AOA by the fractionation process. A higher ratio indicated a higher increase of the AOA by the fractionation process.

7.2.8. Statistical analyses

The data were subjected to the analysis of variance depending on the case according to Kuehl (1999). The SAS (Statistical Analysis System, v. 8.1; SAS Institute Inc., Cary, NC, USA) software was used. Treatment effects were evaluated by F test at $\alpha=0.05$. The results were reported as mean±SE for individual phenolics. Means comparison was done by Duncan multiple range test at α level= 0.05. The relationship among the different variables was measured by the Pearson correlation coefficient (R). Multiple and simple linear regression analyses were carried out to model the existing relationships among individual phenolics and the AOA.

7.3. Results

7.3.1. Phenolic profile of TNG73

The comparison of peaks' retention times of the standar solutions (Figure 25A) with those of sample fractions (Figure 25B-D) show that three main phenolic compounds, namely rutin, catechin, and gallic acid were encountered in TNG73. Rutin was encountered mainly in free form as indicated by its predominance in DE fraction (Figure 25B). Gallic acid and catechin were found in bound forms and were released after acid and base treatments, respectively. Therefore, gallic acid was the main phenolic compound in the AHPA fraction (Figure 25C), while catechin predominated in the BHPA fraction (Figure 25D).

Results of quantitative evaluation of the phenolic compounds are presented in Table 37 for the different fractions of both the skin and the flesh and both the LLP and PHM methods. The catechin content amounted to 11.68 mg/kg sample dry matter in the DE fraction of the flesh and 268.24 mg/kg sample dry matter in BHPA fraction of the skin. The gallic acid content ranged from 2.48 mg/kg sample dry matter in the FPA fraction of the flesh to 25.56 mg/kg sample dry matter in AHPA fraction of the skin. The highest rutin content was observed in the DE fraction of the flesh (67.06 mg/kg sample dry matter), while the lowest value was observed in the AHPA fraction of the skin (9.91 mg/kg sample dry matter). The results showed that catechin was the major phenolic compound in TNG73 with an average of 72.15 mg/kg sample dry matter. It was followed by rutin, and then gallic acid averaging 31.09 and 7.09 mg/kg sample dry matter, respectively.

7.3.2. Distribution of rutin, gallic acid, and catechin in the skin and the flesh of TNG73

The contents of catechin and gallic acid were higher in the skin in comparison to the flesh (Figure 26). However, rutin was encountered at higher concentration in the flesh than the skin. These results are in agreement with Kondo *et al.* (2002) who reported total phenolics content higher in the skin than the flesh of fruits. Similarly, Duvivier *et al.* (2008b) found higher content of total phenolics in the skin of TNG73 in comparison to the flesh.

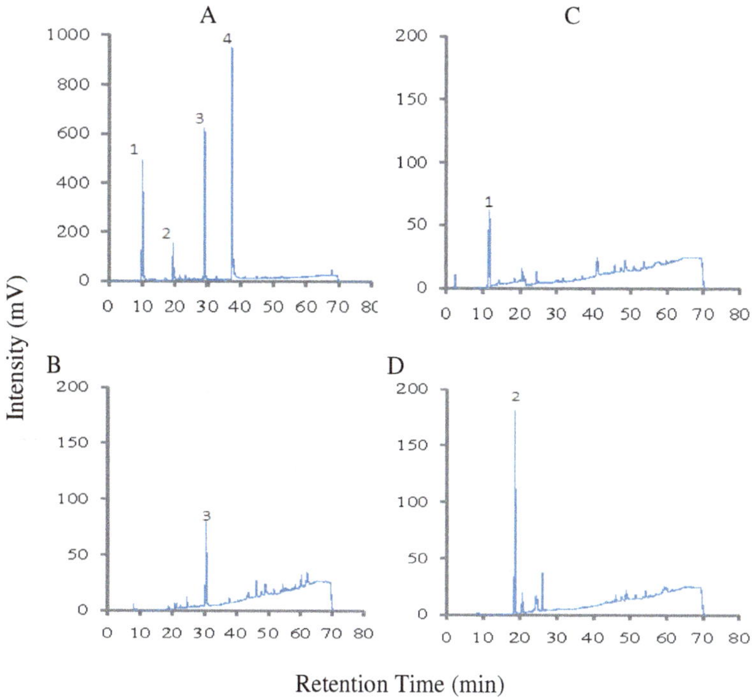

Figure 25. Representative chromatograms of a standard phenolics solution and fractions of TNG73.

A: standard phnolics, B: DE fraction, C: AHPA fraction, D: BHPA fraction; 1: gallic acid, 2: catechin, 3: rutin, 4: quercetin.

Table 37. Catechin, gallic acid, and rutin contents in different fractions of TNG73

A[1]	B[2]	C[3]	(Mean ±SE, n=2)[4]		
			Catechin	Rutin	Gallic acid
Flesh	PHM	AHPA	17.68±6.38[de]	10.35±2.82[e]	3.02±0.84[d]
Flesh	PHM	BHPA	246.61±1.22[b]	11.68±4.09[e]	3.76±0.48[d]
Flesh	LLP	DE	11.68±2.92[e]	67.06±1.75[a]	4.29±1.19[cd]
Flesh	LLP	EA	17.41±0.10[de]	38.23±2.19[c]	4.83±1.78[cd]
Flesh	PHM	FPA	15.77±7.31[de]	41.80±4.36[c]	2.48±0.91[d]
Skin	PHM	AHPA	64.10±0.56[c]	9.91±0.08[e]	25.56±0.28[a]
Skin	PHM	BHPA	268.24±3.65[a]	10.62±0.28[e]	11.47±0.30[b]
Skin	LLP	DE	22.27±6.36[de]	58.85±1.51[b]	4.50±1.39[cd]
Skin	LLP	EA	30.89±0.94[d]	34.44±0.91[cd]	7.16±0.04[c]
Skin	PHM	FPA	26.85±5.90[de]	27.97±3.40[d]	3.81±1.05[d]
Mean			72.15±0.88	31.09±0.48	7.09±0.17

[1]A: part of the root; [2]B: partition method, LLP: liquid-liquid partition, PHM: pH modification; [3]C: fraction, AHPA: acid-hydrolysable phenolics, BHPA: base-hydrolysable phenolics, DE: diethyl ether, EC: ethyl acetate, FPA: free phenolics; [4]means with different letters in a column are significantly different ($p<0.05$); the results are in mg/kg dry matter.

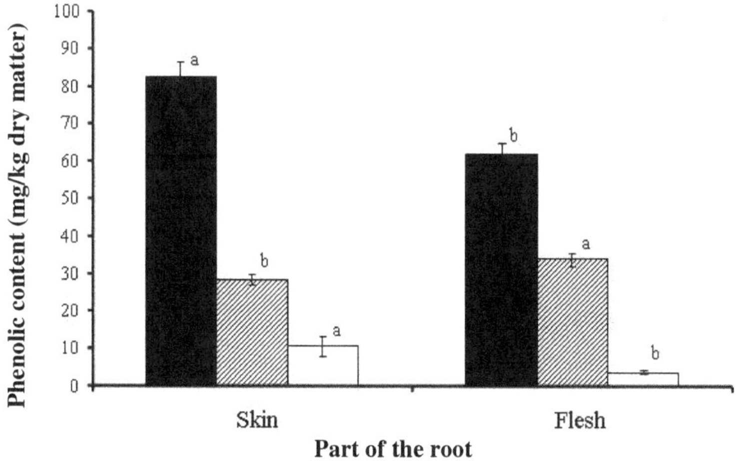

Figure 26. Comparative contents of catechin, rutin, and gallic acid in the skin and the flesh of TNG73.

n : catechin, ▨: rutin, □: gallic acid; the results are mean±SE; means with different letters across samples are significantly different ($p<0.05$).

7.3.3. Comparison of LLP and PHM methods to extract catechin, rutin, and gallic acid of TNG73

When the partition was done by LLP, the amount of catechin and gallic acid extracted was significantly lower in comparison to the PHM method, but the rutin amount was significantly higher (Figure 27). These results suggested that LLP was a better method to extract rutin in comparison to PHM, whereas the latter was better to extract catechin and gallic acid.

7.3.4. Distribution of rutin, catechin, and gallic acid in different fractions of TNG73

The results showed that catechin predominated in BHPA fraction, followed by AHPA, and then EA, DE, and FPA ones (Figure 28). The highest concentration of rutin was found in the DE fraction, followed by EA and FPA, and then AHPA and BHPA ones. Gallic acid predominated in AHPA, followed by BHPA and EA, then DE, and then FPA fractions. The AHPA fraction had the highest content of gallic acid, followed by BHPA, and EA, while FPA fraction exhibited the lowest content. These results suggested that alkaline treatment was more appropriate to extract catechin, while acid treatment was more suitable to extract gallic acid. Liquid-liquid partition between water and DE was a good method to separate rutin from ETOH extract of phenolics.

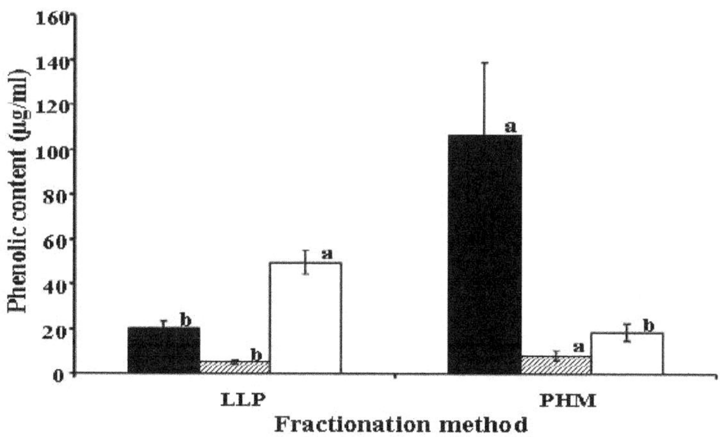

Figure 27. Comparative contents of catechin, rutin, and gallic acid, obtained by liquid-liquid partition (LLP) and pH modification (PHM) methods in TNG73.

n : Catechin, ▨: rutin, □: gallic acid; the results are mean ±SE; means with different letters across methods are significantly different ($p<0.05$).

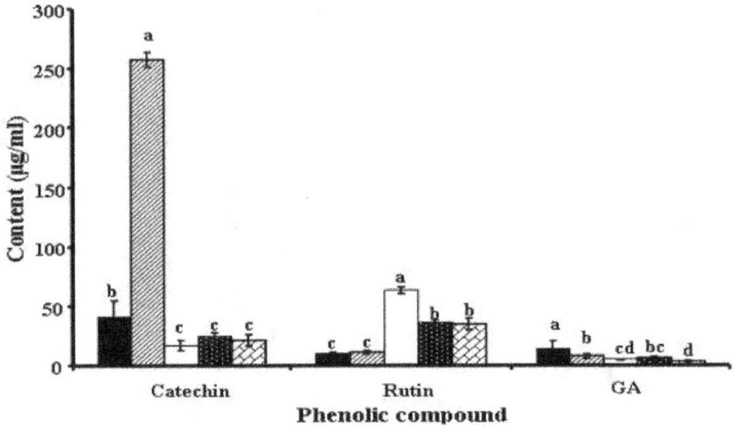

Figure 28. Contents of catechin, rutin, and gallic acid of different fractions of TNG73.

■ : acid-hydrolysable phenolic fraction (AHPA), ▨ : base-hydrolysable phenolic fraction (BHPA), □: diethyl ether fraction (DE), ▦ : ethyl acetate fraction (EA), ▧: free phenolic fraction (FPA); results are mean±SE, n=4; means with different letters within compounds are significantly different ($p<0.05$).

7.3.5. DPPH FRSA of different fractions of TNG73

The DPPH free radical scavenging activity values of the different fractions are presented in Figure 29. The DPPH free radical scavenging activity was concentration-dependent within the range from 0.05 to 500 mg/l. The AHPA fraction of the skin and the flesh (Figure 29A), BHPA fraction of the flesh (Figure 29B), and EA fractions of both the skin and the flesh (Figure 29D) exhibited the highest activities at the concentration of 500 mg/l. In the BHPA fractions of the skin, the DPPH free radical scavenging activity increased until the concentration of 250 mg/l from which it stabilized. In the DE (Figure 29C) and FPA (Figure 29E) fractions of both the skin and the flesh, the DPPH free radical scavenging activity increased until the concentration of 125 mg/l, from which it remained stable. The crude extract DPPH FRSA stabilized at concentration of 2500 mg/l (Figure 29F).

No significant difference in terms of DPPH FRSA was observed between the skin and the flesh in EA and DE fractions. In the AHPA, BHPA, and FPA fractions, the skin showed higher DPPH FRSA than the flesh at concentrations 250 to 500, 0.5 to 500, and 50 mg/l, respectively. For the crude extract, the skin showed higher DPPH FRSA than the flesh at concentrations of 39.06 to 1250 mg/l (Figure 29F). The DE fractions of both the skin and the flesh and FPA fraction of the skin exhibited the lowest DPPH IC_{50}, that is the highest AOA, as indicated in Figure 30. They were followed by the FPA fraction of the flesh, BHPA fraction of the skin, then the EA fraction of both the skin and the flesh, and then the AHPA fraction of the skin. The AHPA and BHPA fractions of the flesh showed the lowest activities (higher DPPH IC_{50}).

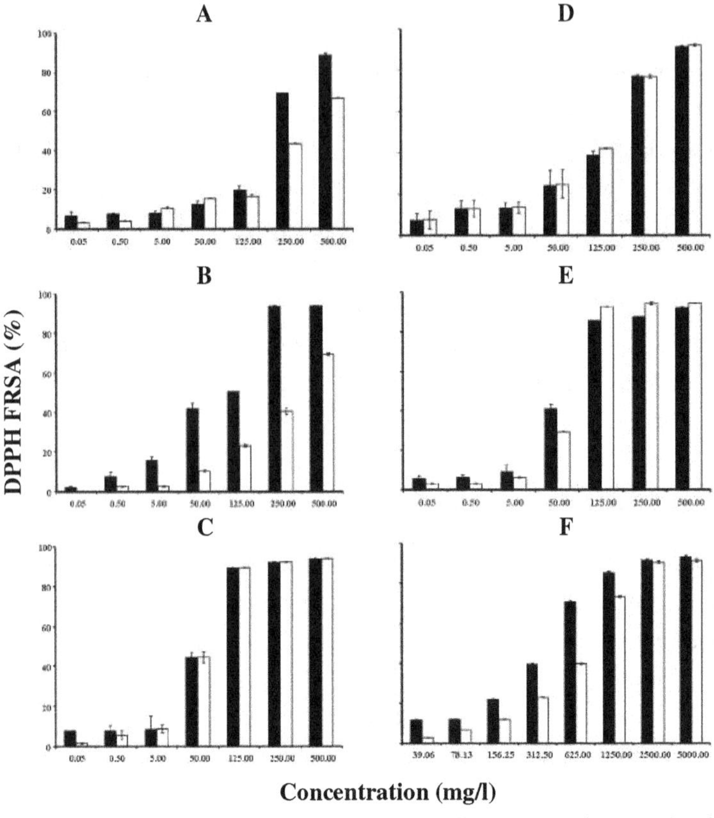

Figure 29. DPPH FRSA vs concentration of different fractions of the skin and the flesh of TNG73.

n : skin, □: flesh; A: acid-hydrolysable phenolic fraction, B: base-hydrolysable phenolic fraction, C : diethyl ether fraction, D: ethyl acetate fraction, E: free phenolic fraction, and F: aqueous ethanol extract. The results are mean±SE, n=3.

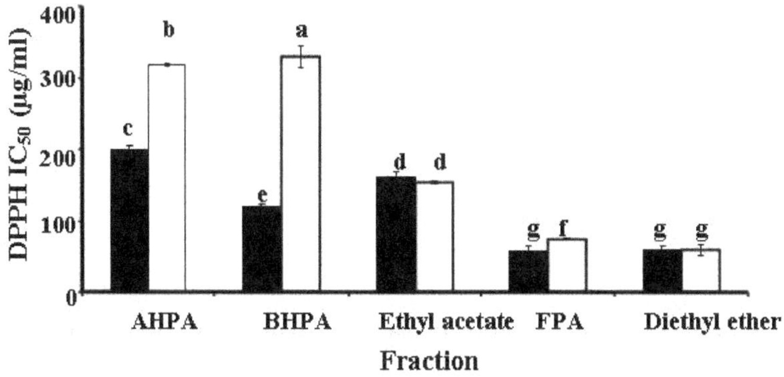

Figure 30. DPPH IC_{50} of different fractions of the skin and the flesh of TNG73. n : skin, □: flesh; AHPA: acid-hydrolysable phenolic fraction, BHPA: base-hydrolysable phenolic fraction, DE : diethyl ether fraction, EA: ethyl acetate fraction, FPA: free phenolic fraction. The results are mean±SE, n=3. Means with different letters within and across fractions are significantly different ($p<0.05$).

The DPPH IC_{50} values of the ethanol crude extract were 415.36 and 816.62 mg/l for the skin and the flesh, respectively, averaging 615.99 mg/l (Table 38). The DPPH IC_{50} of the skin fractions varied from 57.61 mg/l for FPA to 199.94 mg/l for AHPA. For the flesh fractions, it varied from 59.15 mg/l for DE to 331.55 mg/l for BHPA. On average, DE and FPA fractions exhibited the lowest DPPH IC_{50}, followed by EA, then BHPA, and then AHPA. The fractionation process consistently resulted in increase of the AOA. The data presented in Table 39 show that the activities of DE, FPA, and EA fractions represented approximately 10, 9, and 4 times, respectively, that of the ethanol crude extract. AHPA and BHPA fractions were approximately 2 and 3 times, respectively, more active than the crude extract.

7.3.6. Contribution of individual phenolic compounds to the AOA of TNG73

The Pearson correlation coefficients listed in Table 40 show significant ($p<0.05$) and negative ($R<0$) pair-wise correlations between the contents of rutin and the DPPH IC_{50} values of the fractions. The rutin content also correlated negatively with catechin and gallic acid contents of the fractions.

The multiple regression parameters were mathematically modeled in the following equation:

$Y= 0.55\ X_1-0.82\ X_2 +14.57\ X_3 +38.07$; $R^2=0.9999$ and CV=1.25%.

In this equation, Y is the AOA of the fraction expressed in DPPH IC_{50}, i e, the fraction concentration (mg/l) needed to produce 50% DPPH radical inhibition in the conditions of the experiment; X_1, X_2, and X_3 are concentrations (mg/l) of catechin, rutin, and gallic acid, respectively.

Table 38. DPPH IC_{50} of fractions and the crude extract of TNG73

Fraction[1]	-------DPPH IC_{50} (mean±SE)[2]-------		
	Skin, n=3	Flesh, n=3	Mean, n=6
AHPA	199.94±4.42^{b}	320.03± 2.20^{c}	259.99± 60.05^{b}
BHPA	120.81±4.05^{d}	331.55±15.44^{b}	226.18±105.37^{c}
DE	59.32±6.40^{e}	59.15± 6.99^{f}	59.23± 0.08^{e}
EA	161.46±6.58^{c}	159.15± 0.92^{d}	160.31± 1.16^{d}
FPA	57.61±7.53^{e}	74.60± 0.87^{e}	66.10± 8.49^{e}
Crude Extract	415.36±6.09^{a}	816.62±12.19^{a}	615.99±200.63^{a}

[1]AHPA: acid-hydrolysable phenolics, BHPA: base-hydrolysable phenolics, DE: diethyl ether, EA: ethyl acetate, FPA: free phenolics. [2]Fraction means with different letters in a column are significantly different ($p<0.05$); the results are in (mg/l).

Table 39. DPPH IC_{50} ratios of crude extract to fractions

Fraction[2]	Crude extract to fraction ratio[2]		
	Skin	Flesh	mean±SE
AHPA	2.08	2.55	2.31±0.24
BHPA	3.44	2.46	2.95±0.49
EA	2.57	5.13	3.85±1.28
FPA	7.21	10.95	9.08±1.87
DE	7.00	13.81	10.40±3.40

[1]AHPA: acid-hydrolysable phenolics, BHPA: base-hydrolysable phenolics, DE: diethyl ether, EA: ethyl acetate, FPA : free phenolics. [1]Crude extract to fraction ratio = DPPH IC_{50} of the crude extract/ DPPH IC_{50} of the fraction;

Table 40. Matrix of Pearson coefficients of correlation (R) of DPPH IC_{50} to catechin, rutin, and gallic acid contents

	DPPH IC_{50}	Catechin	Rutin	Gallic acid
Catechin	0.617^{NS}			
Rutin	- 0.895*	-0.627^{NS}		
Gallic acid	0.894*	0.213^{NS}	-0.706^{NS}	

* Significant correlation ($p<0.05$); ^{NS}non significant correlation ($p>0.05$).

According to this model, when the catechin and gallic acid concentrations kept constant in the test sample, an increase of 1 mg rutin/l was associated with a decrease of the IC_{50} of 0.82 mg dry matter /l fraction. In contrast, 1 mg/l increase in the catechin concentration was associated with an increase of the IC_{50} of 14.57 mg/l when rutin and gallic acid concentrations remained constant. Similarly, when the concentrations of catechin and rutin stayed constant, 1 mg/l increase in the gallic acid concentration was associated with 0.55 mg/l increase in the IC_{50}. The constant 38.07 a little far from zero meant that the different fractions contained other components contributing to their AOA. The coefficients of determination (R^2) was 0.9999 and that of variation 1.25%, which suggested a high goodness of fit and accuracy of the model. As, lower DPPH IC_{50} means higher AOA, the results suggested that rutin was the main contributor to the AOA of TNG73. It was followed by catechin, and then gallic acid.

7.4. Discussion

TNG73 with 318.79 mg gallic acid equivalent /100 g sample fresh matter when extracted with ETOH (Duvivier *et al.*, 2008a) had a phenolic content comparable to that of strawberries (Asami *et al.*, 2003). Catechin, rutin, and gallic acid were its major individual phenolic compounds. Rutin was encountered mainly in free form. Catechin and gallic acid predominated in bound forms and were released after base and acid hydrolyses, respectively. The presence of these three phenolic compounds in TNG73 and its high AOA make it a potential candidate for functional food and an interesting source of antioxidants. Gallic acid was encountered in the lowest concentration in the samples, followed in ascending order by rutin and catechin. However, the regression analysis showed that rutin was the major contributor to the AOA of TNG73.

These results suggested that rutin was a more powerful antioxidant than catechin. Similarly, Arora *et al.* (1998) reported that rutin is a very powerful antioxidant, comparable to quercetin and *tert*-butylhydroquinone (TBHQ). In effect, both catechin and rutin are flavonoids and get the flavilium backbone. The AOA of flavonoids is closely related to their structure and is due to their electron-donating capacity and the level of hydroxyl and methoxy substitution (von Elbe and Schwartz, 1996). Higher number of hydroxyl and/or methoxy groups on the B-ring corresponds to higher AOA (Arora *et al.*, 1998). Catechin and rutin has the same level and pattern of hydroxylation on the B-ring. Rice-Evans *et al.* (1995) reported that quercetin has a higher AOA than catechin, although both of them have identical number of hydroxyl groups in the same positions, because of the presence of the 2-3 double bonds and the 4-oxo function in the C ring of quercetin. Rutin has the same structure and pattern of hydroxylation than quercetin, except the substitution of the hydroxyl group at C-3 position by the disaccharide rutinose, which can explains the higher AOA of rutin in comparison to catechin.

The presence of substantial concentrations of catechin, rutin and gallic acid makes TNG73 a favorite potential candidate for functional food production and a valuable source of antioxidant. Catechin is a flavonoid with well known AOA. Nagao *et al.* (2005) reported that the AOA of green tea is chiefly due to catechin and a daily consumption of tea containing 690 mg catechins for 12 weeks reduced body fat. This finding suggests that the ingestion of TNG73 might be useful in the prevention and improvement of lifestyle-related diseases, mainly obesity.

Rutin is a poly-phenol pertaining to the class of flavonol glycoside. It has demonstrated high scavenging activity against free radicals process. In humans, it attaches to the iron ion Fe^{2+}, preventing it from binding to hydrogen peroxide, which would otherwise create a highly-reactive free radical that may damage cells (Afanas *et al.*, 1989).

Gallic acid is a simple phenolic compound pertaining to the class of benzoic acids. It has a wide range of biological activities, including anti-oxidant, anti-inflammatory, anti-microbial, and anti-cancer activities (Bachrach and Wang, 2002). It is commonly used in the pharmaceutical industry and as a standard in laboratory for determining the phenol content of various analytes by the Folin-Ciocalteau assay; results are reported in gallic acid equivalents. Raina *et al.* (2008) reported that gallic acid was identified as one of the major constituents of grape seed extract showing strong in vitro efficacy against human prostate cancer cells.

7.5. Conclusions

Catechin, gallic acid, and rutin were three major phenolic antioxidant of TNG73. Liquid-liquid partition was a better method to extract gallic acid in comparison to pH modification, whereas the latter was better to extract catechin and rutin. Catechin was encountered in higher concentration with a maximum content extracted of 268.24 mg/kg, followed by rutin, and then gallic acid at 67.06 and 25.56 mg/kg, respectively. The skin exhibited higher content of catechin and gallic acid, but lower content of rutin than the flesh. Rutin was the first major contributor to the AOA of TNG73, followed by catechin, and then gallic acid.

Chapter 8. Evolution of the total phenolic content and antioxidant activity of TNG73 sweet potato (*Ipomoea batatas* Lam.) flour during 12-month storage

Abstract

The evolution of total phenolic content and antioxidant activity (AOA) of TNG73 sweet potato (*Ipomoea batatas* Lam.) flour was studied during 12-month storage at room temperature and 4°C. The total phenolic contents of samples stored at room temperature and 4°C were not significantly different, but the AOA was on average lower at room temperature than at 4°C. Both total phenolics and AOA exhibited a general trend to decrease during storage. The decrease of the total phenolics of the flesh samples was significant after nine-month storage at room temperature and 10-moonth storage at 4°C. For the skin samples, the decrease of total phenolics was significant after 11-months storage at room temperature and 4°C. The decrease of the AOA of the flesh samples was significant after five-month storage at room temperature and 10-moonth storage at 4°C. For the skin samples, the decrease of AOA was significant after 11-months storage at room temperature and 4°C.

Key words: antioxidant activity, storage duration, sweet potato, total phenolics

8.1. Introduction

In 1998, sweet potato was reported by the International Center of Potato as the fifth and seventh most important food crop, on the basis of its volume of production, in developing countries and the world, respectively. Its total production and utilization in the world is estimated at 130 million tons (FAOSTAT, 2005). It is also estimated that 98% of its total production and utilization was in developing countries (Scott, 1992). In 2002, the annual per capita consumption of sweet potato roots was estimated at 43, 7, 2, and 0.1 kg in China, Japan, United States, and Europe, respectively (FAOSTAT, 2005). Sweet potato roots are rich in crude protein, minerals, and carotenoids (Picha, 1985) and are a staple food in many developing countries (CIP, 1998).

Therefore, sweet potato can play an important role in solving problems associated with global food and energy shortage (Kozai *et al.*, 1996), and in enhancing disease prevention and health promotion as some of its cultivars have been reported to have high antioxidant activity (AOA) due to the high content of phenolics (Teow *et al.*, 2007). Thus, it is important to diversify its use and expand its distribution in the world. However, expansion of sweet potato distribution in the world requires reduction of its bulk volume and weight and extension of its storage life to facilitate its shipping. Such requirements can be achieved by drying, but without compromising its nutritional and functional qualities during the drying process and/or the storage.

It has been reported that during processing and storage, the phenolic contents and AOA of fruits and vegetables decreased (Walkowiak-Tomczak, 2007). Dietrich *et al.* (2004) and Borowska *et al.* (2005) suggested that the most valuable sources of natural antioxidants are fresh fruit and vegetables, since procedures, such as fruit comminution, thermal processing, pressing, clarification, filtration, and storage, may considerably lower the AOA of the product in relation to its raw materials. Contrarily, an increase of polyphenol concentration as a result of thermal processing of raw materials; such as, broccoli, peppers, spinach and beans, has also been reported (Turkmen *et al.* 2005). Padda (2006) reported a significant increase in total phenolic content from 55.30 to 125.80 μg chlorogenic acid equivalent /100 g fresh matter of raw roots of sweet potato cultivar Beau-Regard, which had been stored at 5 °C for 4 weeks. Padda (2006) also reported that long term storage for 12 months of raw sweet potatoes at 15 °C resulted in no significant change in total phenolic content. These findings suggested that after an increase in the first month of storage, the total phenolic content decreased in the following months.

Padda (2006) further reported an increase of the AOA of sweet potato roots from 48.50 to 178.40 mg Trolox equivalent/100 g fresh matter during the 4-week storage at 5°C. This increase in the AOA was attributed by the author to the increase in the total phenolic content with a coefficient of correlation of 0.995.

The differences among the findings reported by different researchers suggested that the behavior of the total phenolic content and the AOA of food commodities during storage is genotypically dependant. This genotypic dependency impedes extrapolations, suggesting the need of particular experiment to elucidate the evolution of total phenolic content and AOA of any food crop that is of interest. The objective of this study was to analyze the evolution of the total phenolic content and the AOA of flour of TNG73 during 12-month storage.

8.2. Material and Methods

8.2.1. Materials

Freshly harvested sweet potato tuberous roots purchased from a local market in Pingtung, Taiwan, in September 2007. Samples were kept at -20^{0}C before processing to slow down chemical reactions. DPPH radical, Folin-Ciocalteu phenol's reagent, gallic acid, sodium carbonate (Na_2CO_3), and sodium hydroxide were purchased from Sigma Chemical Co. (St. Louis, MO, USA). All the other reagents were either extra pure or analytical grade.

8.2.2. Experiment design

The study was carried out according to a 2*2*12 factorial experiment in a completely randomized design with three replicates, yielding 48 treatments and 144 experiment units. The first factor consisted of two parts of the root (flesh and skin), the second factor corresponded to two storage conditions (room temperature and 4°C), and the third one referred to 12 storage durations (month). The response variables were the total phenolic content and the AOA.

8.2.3. Sample preparation

The sweet potato samples were randomly assigned to two groups (A and B) of 180 roots. Each group was then sub-divided into 36 sub-groups of five roots. The roots were washed in tap water, left to dry at room temperature, and hand-peeled with a knife. The flesh and skin samples of each sub-group were separately cut into small pieces of 0.5 to 1 cm of thickness, generating 72 batches for each group (36 batches of flesh samples and 36 batches of skin samples). These samples were dried in oven at 75°C for 24 h to reach a water activity of about 0.4 and a water content of about 5%. Each dried sample was then ground into a fine powder using a laboratory mill, and the powder was sealed in a plastic bag. The bags of group A were stored at room temperature in desiccators, whereas the bags of group B were stored at 4°C. The sample dry mater was measured according to the official methods of analysis of the Association of Analytical Chemists (AOAC, 1970).

Water activity of the samples was measured in triplicate in an Aw Quick water activity meter (Rotronic Instrument Corp., Huntington, N.Y., U.S.A.).

8.2.4. Crude extract preparation

Every month, three bags each of the four groups of samples; flesh powder stored at 4°C, skin powder stored at 4°C, flesh powder stored at room temperature, and skin powder stored at room temperature; were extracted. An aliquot of 4 ±0.01 g powder was mixed with 40 ml ethanol in a 250 ml flask. The flask was then shaken at 100 rpm in dark for 24 h in a water bath at room temperature for complete extraction (Huang *et al.*, 2004; Auerbach and Gray, 1999). The mixture was filtered through Whatman no. 1 filter paper (Whatman Inc., Clifton, NJ). The residue was washed with an additional 40 ml ethanol. The filtrate was evaporated to dryness using a rotary evaporator model BUCHI 111 equipped with a water-bath BUCHI 461 (Switzerland) and a vacuum pump EYELA A-3S (Tokyo, Japan). The dried residue, referred to as crude extract, was weighted, re-dissolved in ethanol to a concentration of 10 mg/ml, and stored at -20 °C until analysis.

8.2.5. Total phenolic content measurements

The total phenolic content was assayed according to the method of Folin and Ciocalteu (1927) as reported by Miliauskas *et al.* (2004) using gallic acid as standard. The results were expressed as mg gallic acid equivalent (GAE)/100 g sample dry matter.

8.2.6. Antioxidant activity (AOA) assay

The AOA of nine extract concentrations from 0.04 to 10 mg/ml was measured by DPPH assay, which was conducted according to Molyneux (2004) in three replicates. Absorbencies were monitored at 517 nm (A_{517}) in a spectrophotometer Beckman Coulter DU 730 UV/VIS, and DPPH FRSA was determined using the following formula: DPPH FRSA (%) $=[(A_0-A_f)/A_0]*100$, where A_0 is the blank absorbance and A_f, the test sample absorbance.

The DPPH FRSA was plotted against the extract concentration, and the DPPH IC_{50}, i.e., the extract concentration that produces 50% DPPH inhibition was determined by graphic interpolation.

8.2.7. Statistical analyses

Mean and standard error of the dependant variables, AOA and total phenolics, were computed in Excel. A scatter plot of total phenolic or DPPH IC_{50} against the storage duration was drawn, and the suggested model equation was determined separately for samples stored at room temperature and 4°C. The scatter plot was drawn from the fifth month to have at least 6 points and degrees of freedom for each regression line. The regression parameters: intercept, slope, Pearson coefficient of correlation (R), coefficient of determination (R^2), corresponding to the evolution of the AOA and total phenolic content during 12-month storage were calculated.

The R value indicates the strength of the relationship between the total phenolics or AOA and the storage duration, while R^2 indicates how predictable the total phenolic content and the AOA are during the storage, knowing the initial values (Kuehl, 1999). The relationship between the total phenolics and AOA was also measured by the Pearson correlation coefficient. The trend significance was evaluated by comparing the absolute values of the computed coefficient of correlation ($|R|$) with the reference coefficient of correlation values (ρ) at α level of 0.05 and the corresponding degree of freedom, n-2, where n is the number of months of storage + 1 (starting point before storage). The trend was considered as significant when the $|R|$ was greater than ρ.

8.3. Results

8.3.1. Total phenolic content

The total phenolic content exhibited a general trend of decrease during the storage. Figure 31 shows the trend for the pooled data collected monthly of the flesh and the skin stored at room temperature and 4°C. The data summarized in Figure 32 showed no significant difference between the samples stored at 4°C and those at room temperature. The total phenolic content exhibited a general trend of decrease during the storage for both the flesh and the skin samples stored at room temperature and 4°C. The total phenolic content of the flesh samples, which was 101.31 mg GAE/100g dry matter, decreased after 12 months of storage to 55.98 mg GAE/100g dry matter at room temperature and 73.94 mg GAE/100 g dry matter at 4°C. The skin samples phenolic content, which was 308.09 mg GAE/100 g dry matter before storage, reached after 12 month-storage 159.35 mg GAE/100 g dry matter at room temperature and 191.98 mg GAE /100 g dry matter at 4°C.

However, low correlations were observed between the storage duration and the total phenolic content. The regression parameters presented in Table 41 show the correlations between the total phenolic content the flesh samples and the duration of storage after the samples had been stored for 12 months. The correlation between the total phenolic content of the flesh samples and the duration of storage was significant only from the ninth month of storage at room temperature, or after 10 months of storage at 4°C. For the skin samples, the trend of decrease of the total phenolic content was significant only after 11 months of storage both at room temperature and at 4°C (Table 42).

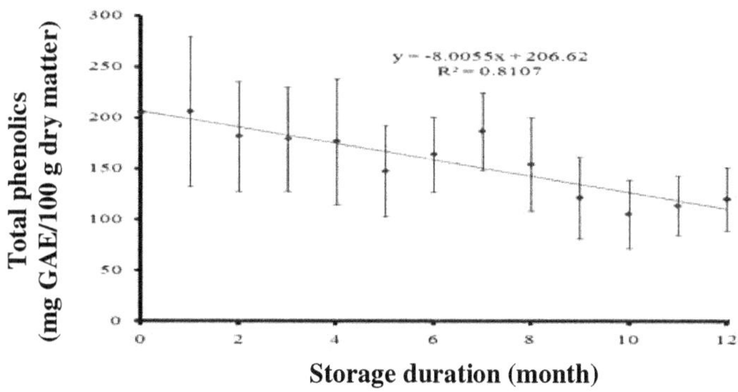

Figure 31. General evolution trend of total phenolics content of TNG73 flour during 12-month storage.

Each point represents the pooled data (mean±SE, n=12) of flesh and the skin stored at room temperature and 4°C.

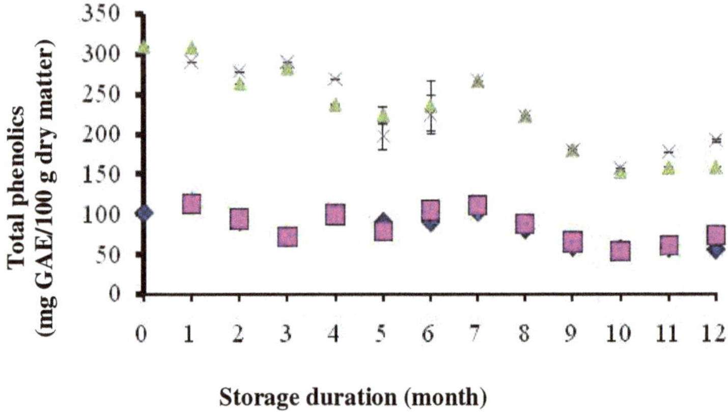

Figure 32. Changes in total phenolics of the flesh and the skin flours of TNG73 during 12-month storage at room temperature and at 4°C.

♦: flesh samples stored at room temperature, ■: flesh samples stored at 4°C, ▲: skin samples stored at room temperature, X: skin samples stored at 4°C. Point values are means of three replicates±SE; when the error bar is not seen, it is covered by the mark.

Table 41. Regression parameters for the evolution of the total phenolic content of TNG73 flesh flour during 12-month storage

Month	A[1]	Regression parameter[†]				
		Intercept	Slop	R^2	$\|R\|$	$\rho_{(\alpha=0.05)}$
4	RT	116.34	-8.26	0.457	0.676^{NS}	0.878
5	RT	113.62	-6.22	0.428	0.654^{NS}	0.811
6	RT	111.18	-4.77	0.386	0.621^{NS}	0.754
7	RT	107.01	-2.68	0.180	0.424^{NS}	0.707
8	RT	107.67	-2.96	0.275	0.524^{NS}	0.666
9	RT	110.78	-4.18	0.455	0.675*	0.632
10	RT	112.55	-4.76	0.578	0.760*	0.602
11	RT	107.02	-4.06	0.573	0.757*	0.576
12	RT	107.30	-4.14	0.638	0.799*	0.553
4	4°C	116.67	-8.39	0.488	0.699^{NS}	0.878
5	4°C	115.83	-7.75	0.585	0.765^{NS}	0.811
6	4°C	105.12	-3.92	0.231	0.481^{NS}	0.754
7	4°C	104.63	-1.51	0.048	0.219^{NS}	0.707
8	4°C	105.27	-1.79	0.090	0.300^{NS}	0.666
9	4°C	108.65	-3.05	0.252	0.502^{NS}	0.632
10	4°C	111.88	-4.13	0.413	0.643*	0.602
11	4°C	108.97	-3.88	0.390	0.625*	0.576
12	4°C	107.06	-3.44	0.387	0.622*	0.553

[1]A: storage condition; RT: room temperature (25°C); [†]R^2: coefficient of determination; R: computed coefficient of correlation; $\rho_{(0.05)}$: reference coefficient of correlation at 5% of probability. *Significant change ($p<0.05$). ^{NS}non significant change ($p>0.05$).

Table 42. Regression parameters for the evolution of the total phenolic content of TNG73 skin flour during 12-month storage

Month	A[1]	Regression parameter[†]				
		Intercept	Slop	R^2	$\|R\|$	$\rho_{(\alpha=0.05)}$
4	RT	298.03	-35.03	0.665	0.815^{NS}	0.878
5	RT	278.78	-20.59	0.402	0.634^{NS}	0.811
6	RT	264.38	-11.95	0.216	0.465^{NS}	0.754
7	RT	250.05	-4.79	0.049	0.221^{NS}	0.707
8	RT	248.17	-3.98	0.048	0.229^{NS}	0.666
9	RT	252.73	-5.69	0.120	0.346^{NS}	0.632
10	RT	258.54	-7.63	0.235	0.485^{NS}	0.602
11	RT	310.33	-13.34	0.834	0.913*	0.576
12	RT	309.22	-13.04	0.858	0.926*	0.553
4	4°C	227.44	-3.98	0.022	0.148^{NS}	0.878
5	4°C	236.93	-3.14	0.021	0.145^{NS}	0.811
6	4°C	235.75	-2.44	0.020	0.141^{NS}	0.754
7	4°C	227.82	-1.53	0.010	0.100^{NS}	0.707
8	4°C	230.53	-0.37	0.001	0.032^{NS}	0.666
9	4°C	298.51	-2.63	0.046	0.214^{NS}	0.632
10	4°C	245.91	-5.09	0.171	0.414^{NS}	0.602
11	4°C	308.62	-12.76	0.743	0.862*	0.576
12	4°C	302.64	-11.38	0.723	0.850*	0.553

[1]A: storage condition; RT: room temperature (25°C); [†]R^2: coefficient of determination; R: computed coefficient of correlation; $\rho_{(0.05)}$: reference coefficient of correlation at 5% of probability. *Significant change ($p<0.05$). [NS]Non significant change ($p>0.05$).

8.3.2. Antioxidant activity

The DPPH FRSA was concentration-dependant. Figure 33 shows a representative graph of DPPH inhibition versus concentration for the extract from the flesh of TNG73 after 1-month storage at room temperature. At concentrations lower than or equal to 0.16 mg/ml, the curve showed a lag phase. Between extract concentrations of 0.16 and 1.25 mg/ml, the DPPH FRSA increased steadily with increasing concentrations. From the concentration of 2.5 mg/ml, the DPPH FRSA tended to stabilize.

This pattern of variation of the DPPH FRSA activity in function of the extract concentration suggested that DPPH IC_{50} was a more reliable parameter than the % DPPH inhibition to evaluate the AOA. The DPPH IC_{50} exhibited a general trend of increase during the storage. Figure 34 shows the trend for the pooled data of the flesh and the skin flours stored at room temperature and 4°C every month.

The data summarized in Figure 35 show that on average, the flesh samples stored at 4°C exhibited lower DPPH IC_{50} (higher AOA) than those stored at room temperature. No significant difference was observed between the skin samples stored at room temperature and at 4°C in terms of DPPH IC_{50}. The DPPH IC_{50} of the extract from the flesh samples, which was 528.65 mg/l before the storage, increased after 12 months of storage to 946.87 mg/l at room temperature and 739.08 mg/l at 4°C. The DPPH IC_{50} of the skin samples, which was 410.96 mg/l before storage, increased after 12 months of storage to 478.74 mg/l at room temperature and 678.56 mg/l at 4°C.

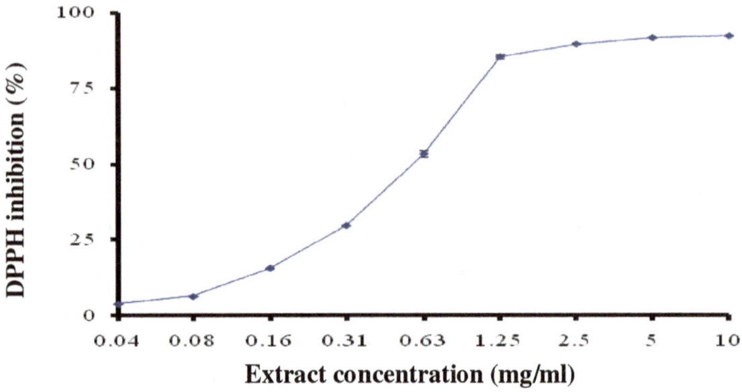

Figure 33. DPPH inhibition vs concentration of the extract from the flour of the flesh of TNG73 after 1 month of storage at room temperature. Results are mean±SE, n=3. When the error bar is not seen, it is covered by the mark.

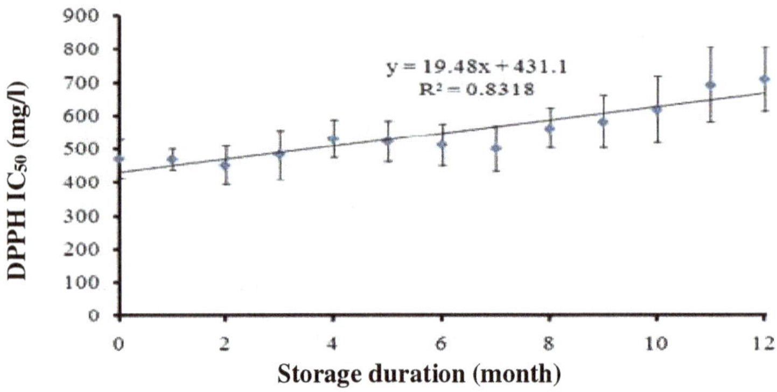

Figure 34. General evolution trend of DPPH IC_{50} of TNG73 flour during 12-month storage.

Each point represents the pooled data (mean±SE, n=12) of the flesh and the skin flours stored at room temperature and 4°C.

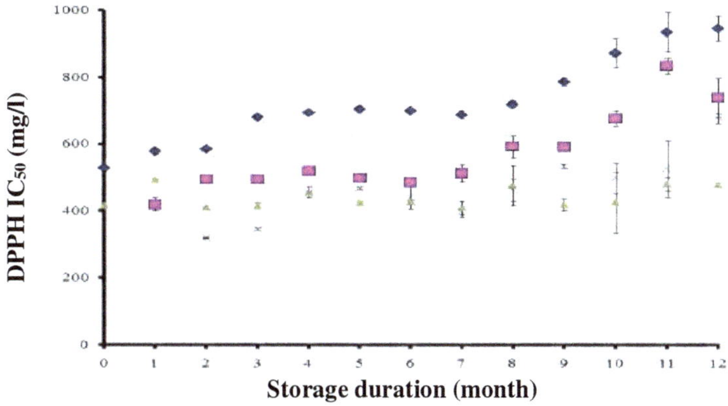

Figure 35. Changes in DPPH IC_{50} of the flesh and the skin flours of TNG73 during 12-month storage at room temperature and at 4°C.
♦: flesh samples stored at room temperature, ■: flesh samples stored at 4°C, ▲: skin samples stored at room temperature, X: skin samples stored at 4°C. Point values are means of three replicates ±SE; when the error bar is not seen, it is covered by the mark.

However, the trend of increase of the DPPH IC_{50} was not always significant. The increase of the DPPH IC_{50}, i.e., the decrease of the AOA, of the flesh samples stored at room temperature was significant after five months (Table 43), while at 4°C, the increases of the DPPH IC_{50} of the samples were significant only after 11-month storage. For the skin samples, the decrease of the DPPH IC_{50} was significant after 11 months of storage both at room temperature and at 4°C (Table 44). The data summarized in Figure 36 indicate that the total phenolic content was highly correlated with the AOA.

Table 43. Regression parameters for the evolution of DPPH IC_{50} of the flour of TNG73 flesh during 12-month storage

Month	A[1]	Regression parameter†				
		Intercept	Slop	R^2	|R|	$\rho_{(\alpha=0.05)}$
4	RT	509.60	42.46	0.585	0.765^{NS}	0.878
5	RT	513.71	39.78	0.676	0.822*	0.811
6	RT	520.75	35.55	0.715	0.846*	0.754
7	RT	535.25	28.31	0.645	0.803*	0.707
8	RT	544.19	24.48	0.636	0.798*	0.666
9	RT	547.19	23.35	0.683	0.827*	0.632
10	RT	546.30	23.65	0.747	0.864*	0.602
11	RT	539.66	30.40	0.880	0.938*	0.576
12	RT	534.54	31.80	0.905	0.951*	0.553
4	4°C	495.91	8.46	0.048	0.219^{NS}	0.878
5	4°C	504.33	-2.15	0.005	0.073^{NS}	0.811
6	4°C	505.39	-1.51	0.004	0.065^{NS}	0.754
7	4°C	513.80	-2.69	0.018	0.134^{NS}	0.707
8	4°C	500.06	3.20	0.028	0.166^{NS}	0.666
9	4°C	487.56	7.87	0.163	0.404^{NS}	0.632
10	4°C	484.12	9.03	0.261	0.511^{NS}	0.602
11	4°C	380.08	29.45	0.727	0.853*	0.576
12	4°C	379.15	29.67	0.779	0.882*	0.553

[1]A: storage condition; RT: room temperature (25°C); †R^2: coefficient of determination; R: computed coefficient of correlation; $\rho_{(0.05)}$: reference coefficient of correlation at 5% of probability. *Significant change ($p<0.05$). ^{NS}Non significant change ($p>0.05$).

Table 44. Regression parameters for the evolution of DPPH IC_{50} of the flour of TNG73 skin during 12-month storage

Month	A[1]	Regression parameter†				
		Intercept	Slop	R^2	|R|	$\rho_{(\alpha=0.05)}$
4	RT	435.28	2.68	0.013	0.112^{NS}	0.878
5	RT	440.45	0.90	0.002	0.049^{NS}	0.811
6	RT	441.42	1.72	0.014	0.117^{NS}	0.754
7	RT	441.30	1.66	0.019	0.138^{NS}	0.707
8	RT	440.07	1.13	0.013	0.113^{NS}	0.666
9	RT	430.07	2.62	0.059	0.242^{NS}	0.632
10	RT	394.29	14.54	0.294	0.542^{NS}	0.602
11	RT	431.33	21.22	0.436	0.660*	0.576
12	RT	427.37	22.31	0.475	0.689*	0.553
4	4°C	391.14	0.04	0.000	0.000^{NS}	0.878
5	4°C	377.19	10.48	0.115	0.339^{NS}	0.811
6	4°C	371.24	14.07	0.265	0.515^{NS}	0.754
7	4°C	386.15	6.62	0.084	0.288^{NS}	0.707
8	4°C	387.70	5.95	0.096	0.310^{NS}	0.666
9	4°C	367.89	13.09	0.275	0.524^{NS}	0.632
10	4°C	374.92	10.74	0.293	0.541^{NS}	0.602
11	4°C	348.22	15.82	0.594	0.771*	0.576
12	4°C	324.81	21.23	0.663	0.814*	0.553

[1]A: storage condition; RT: room temperature (25°C); †R^2: coefficient of determination; R: computed coefficient of correlation; $\rho_{(0.05)}$: reference coefficient of correlation at 5% of probability. *Significant change ($p<0.05$). ^{NS}Non significant change ($p>0.05$).

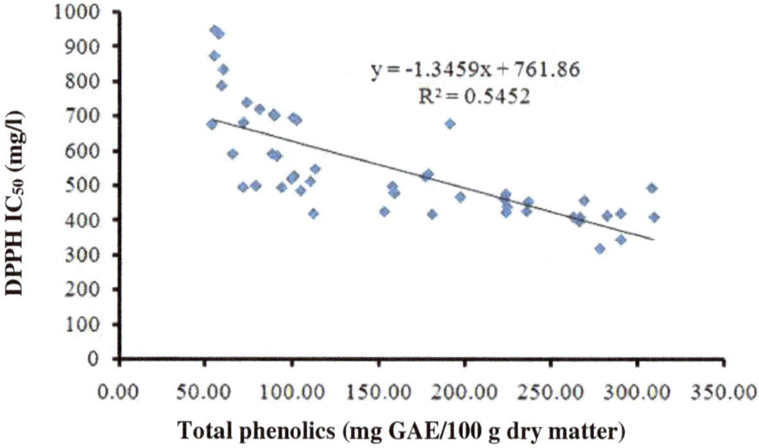

Figure 36. Correlation between total phenolic content and DPPH IC_{50}.

Each point is the mean of three replicates; p=0.0001.

8.4. Discussion

The decrease of the total phenolic content became significant after 9 months of storage for the flesh at room temperature and after 10 months of storage at 4°C. The decrease of the phenolic content of the skin samples was significant only after 11 months of storage both at room temperature and 4°C. These results are in agreement with Padda (2006) in that long term storage for 12 months of raw sweet potatoes cultivar Beau Regard at 15 °C results in no significant change in total phenolic content.

The AOA of the samples was concentration-dependent within the range of 0.04 to 10 mg/ml, as reported by Duvivier *et al.* (2009). Higher AOA and phenolic content were observed for the skin than the flesh. These results are consistent with Kondo *et al.* (2002). As sweet potato skin is discarded as a waste in many cuisines, the high content of phenolic compounds and the high AOA of TNG73 skin make it a potential valuable source of inexpensive antioxidants.

The AOA of the flesh samples started to decrease significantly after 5 months of storage, which is consistent with Walkowiak-Tomczak (2007), Dietrich *et al.* (2004) and Borowska *et al.* (2005). However, the decrease of the AOA of skin samples was not significant during 10-month storage at room temperature. This difference between the skin and the flesh samples in maintaining the AOA during storage was apparently due to the higher content of phenolics in the skin compared to the flesh. This was also evidenced by the high correlation observed between the total phenolics and the DPPH IC_{50}.

The slow decrease of the phenolic content and AOA of TNG73 flour during storage at room temperature and 4°C can be exploited to enhance the usefulness of sweet potato for food security, disease prevention, and health promotion. These results suggested that TNG73 can be processed into flour to reduce costs in distribution and storage and ensure food availability between harvests.

For better retention of the phenolic content and AOA, which are important factors for disease prevention and health promotion, the raw material would better be processed with the skin than without.

8.5. Conclusions

TNG73 is a rich source of phenolic compounds with high AOA, and then is suitable for mass production for disease prevention and health promotion. It is also suitable for processing into flour and storing to ensure food availability between harvests, as its flour maintained high total phenolic content and AOA during long term storage both at room temperature and 4°C. Based on the high AOA and total phenolic content of the skin compared to the flesh, TNG73 would better be processed with the skin than without. In case of use for cooking as a vegetable in areas where sweet potato skin is discarded as a waste, the skin can constitute an inexpensive source of phenolic antioxidants for the food industry.

Chapter 9. Pasting behavior of TNG57, TNG66, and TNG73 sweet potato starch and flour and their applicability in new food product development

Abstract

The pasting behaviors of starch and flour of TNG57, TNG66, and TNG73 sweet potato cultivars were evaluated by using a Rapid Visco Analyzer. The peak viscosity of the starch samples occurred before 5 min around 80°C, and exceeded 495 RVU. Their holding strength was higher than 250 RVU with the final viscosity ranging from 336.00 to 420.72 RVU. The flours' peak viscosity occurred at about 85°C at a maximum time of 7 min and ranged from 61.17 to 80.79 RVU. The starches of the three cultivars had high viscosity, rendering them suitable for processing into food items that require moderate to high gelatinization property of the raw material. The flours had lower gelatinization ability than the starches, but they could still be used as raw materials for new food items, such as cookies and cakes. It is under this premise that this study was conducted.

Key words: sweet potato starch, pasting profile analysis, gelatinization property, new products' development

9.1. Introduction

Sweet potato is an easy-to-grow crop with good adaptability to diverse environmental conditions, high yielding ability, and high energy content (Ravindran *et al.,* 1995). It is rich in dietary fiber, minerals, and vitamins (Woolfe, 1993). It can be grown with no synthetic pesticides and fertilizers (FAO, 1984). It is also suitable for environmentally friendly agriculture and organic food production. Its leaves and roots are edible; thus, it is considered a promising crop for enhancing global food security (Scott, 1992) and minimizing malnourishment problems in developing countries. It can be successfully cultivated throughout tropical and warm temperate regions where there is sufficient water to support its growth. It is estimated that the total acreage for sweet potato production is more than 9 million hectares in more than 100 countries in the world. In terms of fresh matter, sweet potato ranks fifth among the most important food crop after rice, wheat, maize, and cassava in developing countries where 95% of its production, which is estimated at more than 133 million tons annually, is concentrated (Internacional Center of Potato [CIP], 1998).

Some sweet potato cultivars were reported to have anti-carcinogenic properties and/or AOA (Teow *et al.*, 2007). However, neither all food crops, nor all parts of a food crop have equal AOA (Yan, 2006; Rodriguez-Amaya, 2003). With the flesh colors of sweet potato varying from white, yellow to orange and even red among the cultivars, the AOA of its roots may vary over a wide range (Teow *et al.*, 2007). However, in spite of its high nutritional value and its health benefits, sweet potato has been associated with poverty and is considered as "food for the poor", according to a report jointly published by Kenyan Ministry of Agriculture (KMA), and German Technical Cooperation (GTZ) and FAO (1998). This stigma has limited the expansion of its production and consumption in the world.

Only through innovation of the processing technology and consumer education on its nutritional and functional values can it free sweet potato from carrying this stigma; thereby, enhancing a worldwide production and consumption of sweet potato. New and innovative technology in food processing should enhance the development of new sweet potato based food products. This will in turn expand the production and consumption of sweet potato in the world for a better contribution to the global food security. Deep yellow and red colored cultivars are rich in carotenoids and phenolics (Teow *et al.*, 2007), which can be extracted and used as food supplement. Previous studies showed that the ethanol extraction yield of the flesh and skin of TNG57, TNG66, and TNG73 varies between 9.79 and 27.27% (dry matter basis), averaging 14.40±5.88% for raw samples, while for freeze-dried samples, the extraction yield varied between 3.44 to 13.18% (dry matter basis), averaging 6.37±1.66% (Duvivier *et al.*, 2009).

In laboratory experiments, small size samples of raw materials are used; therefore, it is expected to generate small quantity of extraction solid residue. However, if a large quantity of sweet potato raw materials is used, as in a food processing plant, to produce antioxidants, it can be expected that a large volume of the extraction solid residue will be produced, representing over 85% of the raw material on dry matter basis. That extraction solid residue can be used for developing new food items, as it has high starch content. Knowledge about the pasting behaviors of starches and flours is important in the food industry, especially in its product development. Therefore, we designed this study to evaluate the pasting behaviors of sweet potato flour and starch and to analyze the possibility of using the residues, which have usually been treated as a waste after antioxidant extraction, for the development of starch based new food items from sweet potato.

9.2. Materials and methods

9.2.1. Materials

Freshly harvested sweet potato tuberous roots were purchased from a local traditional market in Pingtung, Taiwan, in February 2009.

9.2.2. Flour preparation and starch extraction

The raw sweet potato was cleaned, peeled, and cut into small pieces of 0.5 to 1 cm of thickness. The cut samples were divided into two parts. Of the two parts, one was dried at 75°C for 24 h and milled into fine flour. From the other part, starch was extracted according to Shimelis *et al* (2006) and Kooijman *et al.* (2003). Fresh roots was blended with water (water: sweet potato = 3:1 vol/vol) for 1 min. The mixture was subsequently sieved using 20, 60, and 150-mesh sieves. Fibers were remained on the sieve, while the dispersed starch and the proteins passed through. The crude dispersed starch was again washed three times with water (water: sweet potato = 6:1 vol/vol). The foam on top of the water layer (proteins) was removed. The starch was washed three times with water to remove other contaminants. The wet starch was air-dried for 24 h at 50°C. Figure 37 illustrates the process of extraction.

9.2.3. Pasting behavior evaluation

A Rapid Visco Analyzer (Model: RVA-4, Newport Scientific Pty. Ltd., Sydney, Australia, 1995) was used to evaluate the pasting properties of the flours and starch extracts from sweet potato (Kooljman *et al.*, 2003). A viscogram pasting curve showed the relationship between time, viscosity, and temperature during cooking processes of a sample containing the equivalent of 3 g dry matter (flour or starch) and 25 g distilled water. Test runs were conducted following standard profile 1 which includes: 1 min of mixing-stirring and warming up to 50ºC, 3 min and 42 sec of heating at 12ºC /min up to 95ºC, 2.5 min of holding at 95ºC, 3 min and 48 sec of cooling down to 50ºC, at the same rate as the heating (12ºC/min) and 2 min holding at 50ºC. The process ends after 13 min.

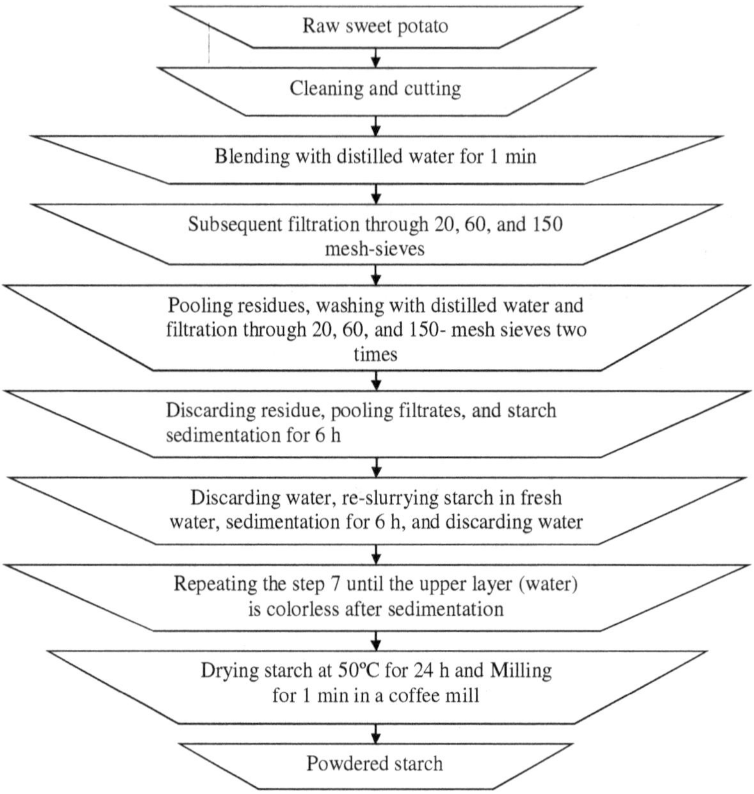

Figure 37. General scheme employed for the extraction of starch from sweet potato.
Source: adapted from Kooljman *et al.* (2003)

Starch gelatinization or pasting curves were recorded on RVA and viscosity was expressed in terms of rapid visco units (RVU) which is equivalent to 10 centipoises. The results of the pasting profile analysis were used to determine the applicability of the extraction solid residue in development of food products that require moderate to high gelatinization ability of the raw materials. Peak viscosity, holding strength or trough, breakdown, final viscosity, and setback from holding strength are the parameters selected for our RVA pasting profile. Peak viscosity refers to the maximum viscosity developed during or soon after the heating portion of the test, while the holding strength or trough is the minimum viscosity after the peak, normally occurring around the commencement of sample cooling. Breakdown is a measure of loss of viscosity on cooling and is computed as peak viscosity minus trough viscosity. The final viscosity is the viscosity at the end of the test, and it is used to compute the setback from the holding strength (final viscosity-holding strength).

9.2.4. Development of new food products

After analyzing the pasting profile of the samples, we developed three different spongy cakes (Figure 38) and three compact cookies (Figure 39) from the flour processed from TNG57, TNG66 and TNG7. Also developed were sweet potato starch puddings (Figure 40). We adapted the general guidelines available for making conventional birthday cakes and cookies from wheat flour to develop the sweet potato spongy cakes. The sweet potato puddings were developed by adapting the recipes for Christmas puddings from rice flour.

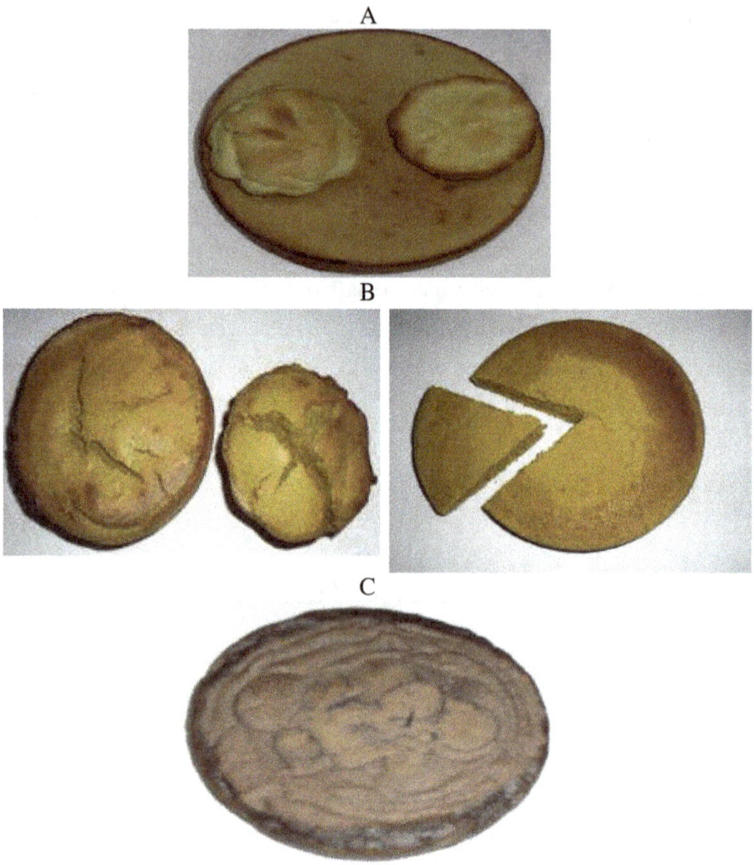

Figure 38. Sponge cakes from sweet potato flour.

A: TNG57, B: TNG66, and C: TNG73.

A

B

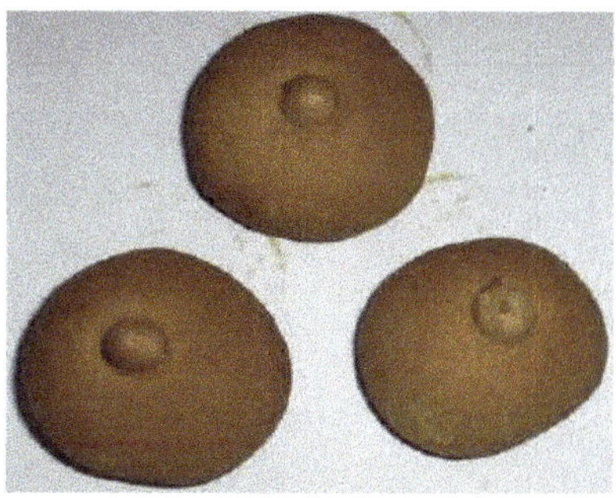

C

Figure 39. Compact cookies from sweet potato flour with peanut top.
A: TNG57, B: TNG66, and C: TNG73.

A

B

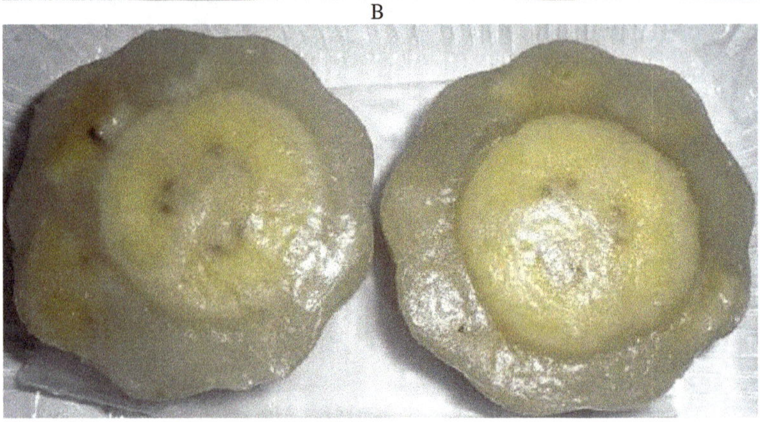

Figure 40. Banana puddings with starch isolated from TNG57 as binder.

A: baked pudding; B: steamed pudding.

9.2.5. Sensory evaluation of the new products

A) Attributes

We carried out a descriptive sensory evaluation of the new products. The first research question for the sensory analysis focused on the overall acceptance of the products on basis of their color, shape, smell, and size. In addition, five specific attributes, namely flavor, taste, sweetness, texture, and hardiness, were evaluated; each attribute was evaluated in a scale of 0 to 10. Each product was given a pair of scores: an actual score and an "optimum score". Figure 41 shows the questionnaires that were used in the sensory evaluation experiment.

B) Definition of optimum score

Overall acceptance and taste are generally known as preference attributes. The tests of preference attributes are known as preference tests, in which the panelists gave the results from "extremely dislike" to extremely like". In the preference tests, the maximum score corresponds to best best score. Flavor, sweetness, texture, and hardiness, are known as intensity attributes. The tests of intensity attributes are known as intensity tests. In intensity tests, the maximum score do not necessarily correspond to the best score. For example, a panelist can find a product "too sweet" or "not sweet enough" according to the level of sweetness he or she would prefer. This level of sweetness that he or she would prefer is his or her ideal sweetness level, which corresponds to a baseline reference score that we called "ideal score" or "optimum score".

The use of the optimum score allowed to generate more information on the products submitted to evaluation. For example, if a panelist evaluated the sweetness of products, A and B, and assigned a score of 5 to A, and a score of 4 to B, those results would only indicate that A is sweeter than B. However, if the panelist also set an optimum score of 4.3 for the sweetness level, the results would also indicate that the panelist found that A was too sweet and B was not sweet enough.

QUESTIONNAIRE FOR SENSORY EVALUATION

PANELIST NAME____________________DATE____________________

Please evaluate all of the below mentioned attributes of these samples. Test the samples in the following order: A B C D E F G

Make notes in reference to your ideal attribute value (I). Example:

Sweetness

0 1 2 3 4 F 5 6 I 7 8 9 10

No sweet — Very sweet

Attribute: overall acceptability

0 1 2 3 4 5 6 7 8 9 10

Extremely dislike — Extremely like

Attribute: flavor

0 1 2 3 4 5 6 7 8 9 10

Less — Extreme

Attribute: taste

0 1 2 3 4 5 6 7 8 9 10

Extremely dislike — Extremely like

Attribute: sweetness

0 1 2 3 4 5 6 7 8 9 10

Not sweet — Very sweet

Attribute: texture

0 1 2 3 4 5 6 7 8 9 10

Very fine — Very gritty

Attribute: softness

0 1 2 3 4 5 6 7 8 9 10

Very hard — Very soft

Figure 41. Questionnaire for sensory evaluation of the new sweet potato products.

When we use the optimum score in a sensory evaluation, the best attribute score is not the highest score, but the score closest to the optimum score.

C) Panelists

The panel consisted of DTAIC and/or NPUST students who were enrolled in Master's program or are pursuing Ph. D. degree. They have all taken the course of food processing in which sensory evaluation of foods was taught. In addition, the panelists have gone through specific training for this experiment. As we used only trained panelists, a restricted number of replicates (n=10) was used.

D) Statistical analysis

We computed in excel the mean and standard error of the scores of each attribute for each product. In a first analysis, each product was evaluated for its own overall acceptance and specific values, as compared to the optimum scores indicated by the panelists. In this case, a t-test of paired scores was carried out at 5% of probability. When *p*-value was lower than 0.05, the actual score of the sample was considered as significantly different from the score of the ideal product of the panel.

In a second analysis, the data were subjected to the analysis of variance of a completely randomized design according to Kuehl (1999). We run a multiple comparison using the Duncan's multiple range test, and we ranked the products on basis of their actual scores and the distance from the optimum scores for each attribute. The critical probability value was set at $\alpha = 0.05$, and a *p*-value lower than 0.05 indicated a significant difference.

9.3. Results

9.3.1. Pasting behavior of starches extracted from TNG57, TNG66, and TNG73

The gelatinization characteristics of the starches of sweet potato cultivars TNG57, TNG66, and TNG73 are listed in Table 45. The results of peak viscosity varied from 498.36 RVU for TNG73 to 539.89 RVU for TNG66, while it reached 613.87 RVU for TNG57. Although TNG66 exhibited the highest RVU value, these results showed that the starches of TNG57, TNG66 and TNG73 all had high viscosity. In terms of the holding strength, there was no significant difference between TNG66 and TNG73; but TNG57 had the lowest among the three cultivars. When comparing with TNG73, TNG66, though had a higher peak viscosity, showed a higher breakdown too. Again, among the three cultivars, TNG57 had the highest breakdown value.

In terms of the final viscosity, the highest was observed in TNG66, followed by TNG73 and TNG57. The gelatinization time varied from 4.13 min for TNG57 to 4.82 min for TNG66, averaging 4.54 min for all three cultivars. Statistically, TNG57 was the fastest to gelatinize, followed by TNG73 and TNG66; however, the range of variation of viscosity peak time from 4.13 to 4.82 min appeared to be of little practical significance. The gelatinization occurred just above 80°C, from 80.23°C for TNG57 to 80.72°C for both TNG66 and TNG73. Figure 42 shows representative RVA viscosity and temperature profiles of starches extracted from TNG57, TNG66, and TNG73.

Table 45. Gelatinization characteristics of starches of TNG57, TNG66, and TNG73

Parameter	(Mean±SE, n=3)[1]		
	TNG57	TNG66	TNG73
Peak viscosity (RVU)	511.56±11.70[b]	539.89±5.36[a]	498.36±7.63[b]
Holding strength (RVU)	262.53±1.50[b]	332.64±3.37[a]	335.92±2.32[a]
Breakdown (RVU)	249.03±10.28[a]	207.25±4.73[b]	162.44±5.34[c]
Final viscosity (RVU)	336.00±0.93[c]	420.72±6.66[a]	403.75±5.78[b]
Setback (RVU)	73.47±0.57[b]	88.08±3.32[a]	67.83±4.72[b]
Peak time (min)	4.13±0.00[c]	4.82±0.02[a]	4.69±0.02[b]
Pasting temperature (°C)	80.23±0.28[c]	80.72±0.04[b]	80.72±0.02[b]

[1]Means with different letters in rows are significantly different ($p<0.05$)

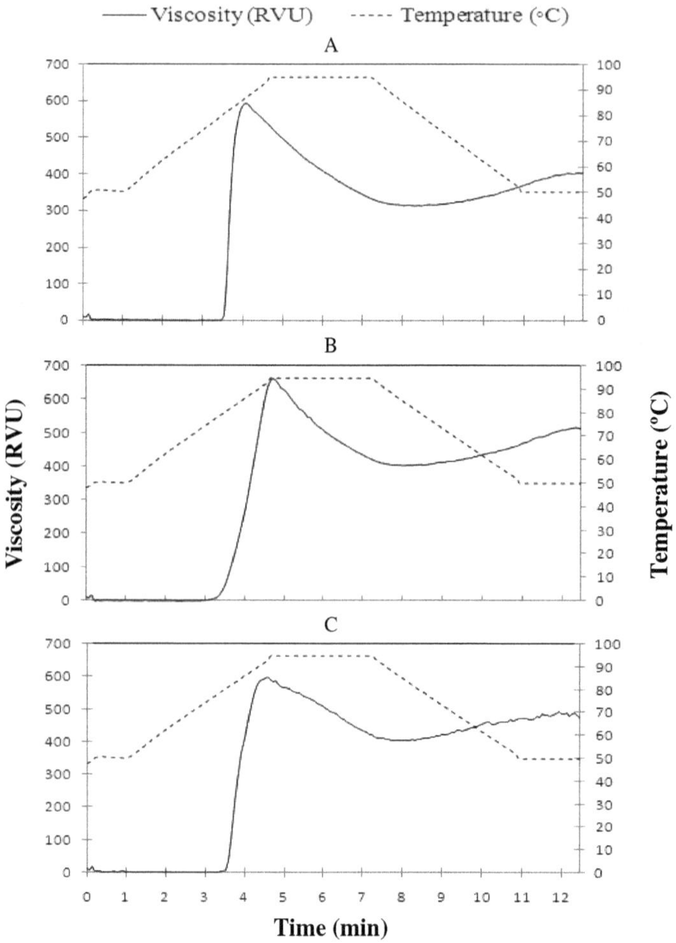

Figure 42. Representative viscograms of starches extracted from sweet potato cultivars.
A: TNG57, B: TNG66, C: TNG73.

9.3.2. Pasting behavior of flour of TNG57, TNG66, and TNG73

The gelatinization characteristics of the flours of TNG57, TNG66, and TNG73 are shown in Table 46. The results of peak viscosity varied from 61.17 RVU for TNG73 to 80.70 RVU for TNG66, while it reached 71.69 RVU for TNG57. The holding strength ranged from 58.54 RVU for TNG73 to 74.04 RVU for TNG66. The flours of the three cultivars were not significantly different in their peak viscosity and holding strength. TNG66 showed the highest breakdown value, while TNG57 and TNG73 were comparable on the values of the breakdown. The three cultivars also exhibited similar final viscosity (93.00 to 105.78 RVU) and setback (30.21 to 36.33 RVU). TNG66 flour showed the lowest peak time (4.70 min), while TNG57 and TNG73 exhibited comparable pasting behavour at 6.71 and 7.00 min, respectively.

The flour samples exhibited significantly lower gelatinization ability than the starch samples. Different shapes in the viscogram were also observed between the starch and flour samples, as shown in Figure 43. The final viscosity of the flour after cooling to 50°C was higher than the peak viscosity during the heating process at 95°C. In contrast, the final viscosity of the starch after cooling to 50°C was higher than the peak viscosity obtained during the heating process at 95°C. This difference in viscogram shapes between starch and flour may be due to the presence of non-starch components in the flour. Higley *et al.* (2003) reported that starch swelling, gelatinization, and retrogradation behavior is dramatically influenced by other constituents within potato tissue. The gelatinization temperatures obtained for both the starches (80.23 to 80.72°C) and the flours (85.23 to 85.72°C) were substantially higher than that reported for wheat starch (55.6 to 63.0°C) (Lineback and Ke, 1975) and for cassava starch (72.7°C) (Zaidul *et al.,* 2007). However, both the starches and the flours of TNG57, TNG66, and TNG73 were suitable for development into new food products.

Table 46. Gelatinization characteristics of flours of TNG57, TNG66, and TNG73

	(Mean±SE, n=3)[1]		
Parameter	TNG57	TNG66	TNG73
Peak viscosity (RVU)	71.69±1.32^{a}	80.79±8.61^{a}	61.17±0.20^{a}
Holding strength (RVU)	69.44±1.49^{a}	74.04±7.99^{a}	58.54±0.44^{a}
Breakdown (RVU)	2.25±0.22^{b}	6.75±0.61^{a}	2.63±0.24^{b}
Final viscosity (RVU)	105.78±2.54^{a}	104.25±11.09^{a}	93.00±0.20^{a}
Setback (RVU)	36.33±1.14^{a}	30.21±3.10^{a}	34.46±0.24^{a}
Peak Time (min)	6.71±0.12^{a}	4.70±0.03^{b}	7.00±0.00^{a}
Pasting temperature (ºC)	85.28±0.50^{a}	85.76±0.08^{a}	85.23±0.02^{a}

[1]Mean with different letters in rows are significantly different (p<0.05).

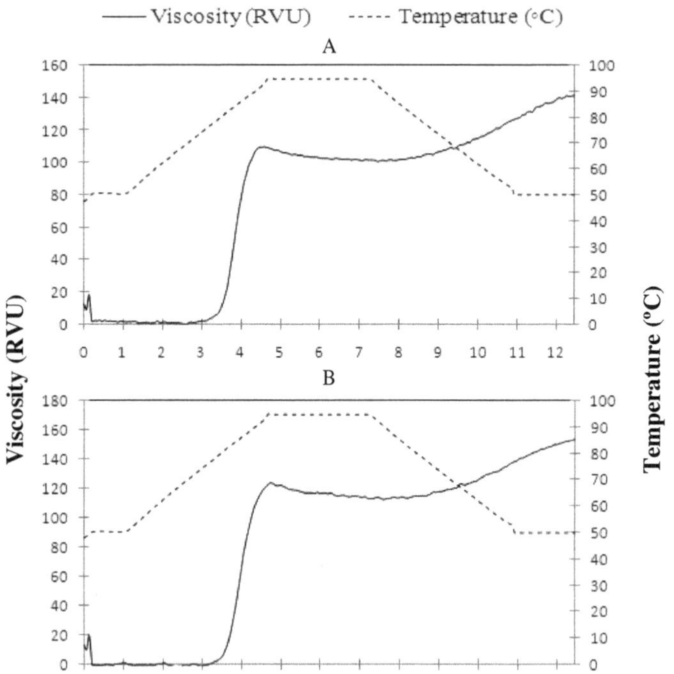

C

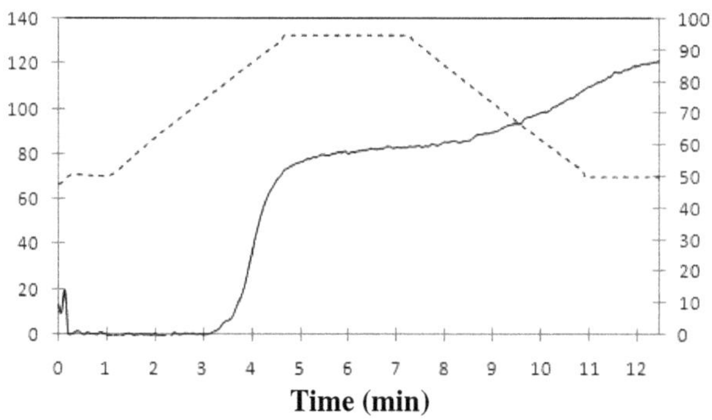

Figure 43. Representative viscograms of flours of sweet potato cultivars.
A: TNG57, B: TNG66, C: TNG73.

9.3.3. Results of the sensory evaluation of the new products

A) Overall acceptance

The data summarized in Table 47 showed that all the products attained a level of acceptability from 5.33 to 7.10 in a scale of 0 to 10. No significant difference was observed among the seven products in terms of overall acceptance. Although this range of acceptability may be considered as quite good, the distance from the optimum score was significant for all the products. Therefore, analyses of specific attributes were conducted to elucidate the point to improve to make the products fully accepted in case of commercialization.

B) Taste

The products' taste score levels ranged from 6.11 (compact cookie from TNG57 flour) to 8.29 (pudding from TNG57 starch) in a scale of 0 to 10 (Table 48). Only the difference between compact cookie from TNG57 flour and pudding from TNG57 starch was significant ($p<0.05$) according to the Duncan's multiple range test. However, the significant differences of the tested product scores from the optimum score indicated that the taste can still be improved.

C) Flavor

Results of the flavor score of the different products are summarized in Table 49. The ideal flavor levels set by the panelists were 3.88 for the spongy cakes, 4.38 for the cookies, and 3.38 for the pudding. The three cultivars were not significantly different in terms of the flavor score. The spongy cakes received on average higher flavor score (5.93±0.47) than the cookies (4.47±0.69). However, higher score did not correspond to better score. Spongy cakes from TNG57 and TNG66 exhibited significant distances from the ideal, while the other products did not. In fact, spongy cakes from TNG57 and TNG66 attained significantly higher flavor scores than the ideal of the panelists.

Table 47. Overall acceptance score of the different sweet potato products developed in the framework of this study

Product[1]	(Mean±SE, n=10)[2]	
	Product score	Distance from the optimum score
A	6.17 ± 0.71[a]	-3.83 ± 0.90[a*]
B	6.08 ± 0.56[a]	-3.92 ± 0.80[a*]
C	6.67 ± 0.76[a]	-3.33 ± 0.77[a*]
D	5.98 ± 0.78[a]	-4.02 ± 0.90[a*]
E	5.33 ± 0.63[a]	-4.67 ± 0.70[a*]
F	5.46 ± 0.90[a]	-4.54 ± 0.89[a*]
G	7.10 ± 0.55[a]	-2.90 ± 0.65[a*]

[1]A, B, C: spongy cakes from TNG57, TNG66, and TNG73, respectively; D, E, F: compact cookies from TNG57, TNG66, and TNG73, respectively. G: pudding from TNG57 starch. [2]Means with a same letter in columns are not significantly different ($p>0.05$); optimum score= 10. *Significant distance from the optimum score.

Table 48. Taste score of the different sweet potato products developed in the framework of this study

Product[1]	(Mean±SE, n=10)[2]	
	Product score	Distance from the optimum score
A	7.04± 0.51[ab]	-2.96 ± 0.71[a*]
B	7.02± 0.56[ab]	-2.98 ± 0.80[a*]
C	7.29± 0.70[ab]	-2.71 ± 0.60 [a*]
D	6.11± 0.59[b]	-3.89 ± 0.61[a*]
E	6.46± 0.54[ab]	-3.54 ± 0.71[a*]
F	6.29± 0.77[b]	-3.71 ± 0.94[a*]
G	8.29± 0.60[a]	-1.71 ± 0.72[a*]

[1]A, B, C: spongy cakes from TNG57, TNG66, and TNG73, respectively; D, E, F: compact cookies from TNG57, TNG66, and TNG73, respectively. G: pudding from TNG57 starch. [2]Means with a same letter in columns are not significantly different ($p>0.05$); optimum score= 10. *Significant distance from the optimum score.

Table 49. Flavor score of the different sweet potato products developed in the framework of this study

Product[1]	(Mean±SE, n=10)[2]		
	Product score	Optimum score	Distance from ideal
A	5.50 ± 0.52^{ab}	3.88 ± 0.45	1.63 ± 0.40^{ab}*
B	6.88 ± 0.61^{a}	3.88 ± 0.45	3.01 ± 0.99^{a}*
C	5.42 ± 0.70^{ab}	3.88 ± 0.45	1.54 ± 0.77^{ab}
D	3.58 ± 0.70^{b}	4.38 ± 0.51	- 0.79 ± 0.50^{b}
E	5.83 ± 0.99^{ab}	4.38 ± 0.51	1.46 ± 1.28 ^{ab}
F	4.00 ± 0.84^{b}	4.38 ± 0.51	- 0.38 ± 0.72^{b}
G	3.67 ± 0.75^{b}	3.38 ± 0.52	0.29 ± 0.38^{b}

[1]A, B, C: spongy cakes from TNG57, TNG66, and TNG73, respectively; D, E, F: compact cookies from TNG57, TNG66, and TNG73, respectively. G: pudding from TNG57 starch. [2]Means with a same letter in columns are not significantly different (p>0.05). *Significant distance from the optimum score.

In terms of flavor, with an average distance of 0.76±0.78, the cookies' scores were closer to the optimum score than the spongy cakes score (average distance=2.06±0.48) and the pudding score (average distance=1.04±0.53) were.

D) Sweetness

Results of the sweetness scores of the different products are summarized in Table 50. The average optimum score levels set by the panelists were 6.27±0.07 for the spongy cakes, 6.63±0.00 for the cookies, and 6.88±0.33 for the pudding. The cookies' scores of the three sweet potato cultivars were not significantly different in terms of sweetness. Although the same procedure was used to make the spongy cakes, the spongy cakes from TNG73 were sweeter, and more appreciated by the panelists for its sweetness than those from TNG57 and TNG66. The sweetness scores of the cookies from the three cultivars were not significantly different from the optimum scores. Similarly, the sweetness score of the pudding was not significantly different from that of the optimum score set by the panelists.

E) Texture

The texture scores of the products are summarized in Table 51. The ideal texture scores set by the panelists were 3.75±0.00 for the spongy cakes, 5.25±0.00 for the cookies, and 3.38±0.54 for the pudding. The three cultivars were not significantly different in terms of texture score of the cookies. However, the spongy cakes from TNG57 were significantly finer in texture than that from TNG73. The spongy cakes received, on average, lower texture score (3.98±0.67) than the cookies (7.10±0.22). However, neither the spongy cakes nor the cookies had significantly different product scores from the optimum scores set by the panelists. The texture score of the puddings was significantly different from that of the optimum score set by the panelists.

Table 50. Sweetness score of the different sweet potato products developed in the framework of this study

Product[1]	(Mean±SE, n=10)[2]		
	Product score	Optimum score	Distance from ideal
A	3.67± 0.48^{b}	6.13±0.71	-2.46±0.50^{b*}
B	3.95± 0.37^{b}	6.29±0.60	-2.34±0.58^{b*}
C	6.17± 0.57^{a}	6.38±0.60	-0.21±0.37^{a}
D	7.25±0.64^{a}	6.63±0.35	0.63 ±0.56^{a}
E	6.75±0.62^{a}	6.63±0.35	0.13 ±0.56^{a}
F	6.67±0.71^{a}	6.63±0.35	0.04 ±0.81^{a}
G	6.67±0.48^{a}	6.88±0.33	-0.21 ±0.35^{a}

[1]A, B, C: spongy cakes from TNG57, TNG66, and TNG73, respectively; D, E, F: compact cookies from TNG57, TNG66, and TNG73, respectively. G: pudding from TNG57 starch.

[2]Means with a same letter in columns are not significantly different (p>0.05). *Significant distance from the optimum score.

Table 51. Texture score of the different sweet potato products developed in the framework of this study

Product[1]	(Mean±SE, n=10)[2]		
	Product score	Optimum score	Distance from ideal
A	3.00±0.65[d]	3.75±0.66	- 0.75±1.04[a]
B	3.68±0.65[cd]	3.75±0.66	- 0.07±1.05[a]
C	5.25±0.44[bc]	3.75±0.66	1.50±0.84[a]
D	6.73±0.79[ab]	5.25 ± 0.58	1.48±1.19[a]
E	7.07±0.45[a]	5.25 ± 0.58	1.82±0.89[a]
F	7.50±0.55[a]	5.25 ± 0.58	2.25±1.00[a]
G	2.67±0.48[d]	3.38 ± 0.54	- 0.71±0.26[a]*

[1]A, B, C: spongy cakes from TNG57, TNG66, and TNG73, respectively; D, E, F: compact cookies from TNG57, TNG66, and TNG73, respectively. G: pudding from TNG57 starch. [2]Means with a same letter in columns are not significantly different ($p>0.05$). *Significant distance from the optimum score.

F) Softness

The softness scores of the products summarized in Table 52 showed that the ideal softness levels set by the panelists were 6.25±0.00 for the spongy cakes and cookies and 5.38±0.57 for the pudding. The three cultivars were not significantly different in the softness scores withing the spongy cakes and within cookies. However, the spongy cakes and the puddings were significantly softer than the cookies. The differences between the product scores and the optimum scores on the softness were not significant for the spongy cakes and the puddings, but the panelists found the cookies too hard (softness score=4.14).

Table 52. Softness score of the different sweet potato products developed in the framework of this study

Product[2]	(Mean±SE, n=10)[2]		
	Product score	Optimum score	Distance from optimum score
A	7.00±0.59[a]	6.25±0.76	0.75±1.10[a]
B	7.17±0.53[a]	6.25±0.76	0.92±1.12[a]
C	5.83±0.38[ab]	6.25±0.76	- 0.42±1.07[ab]
D	4.08±0.78[b]	6.25± 0.27	- 2.17±0.66[b]*
E	4.17±0.67[b]	6.25± 0.27	- 2.08±0.51[b]*
F	4.17±0.71[b]	6.25± 0.27	- 2.08±0.59[b]*
G	6.75±0.74[a]	5.38±0.57	1.38±0.61[a]

[1]A, B, C: spongy cakes from TNG57, TNG66, and TNG73, respectively; D, E, F: compact cookies from TNG57, TNG66, and TNG73, respectively. G: pudding from TNG57 starch.
[2]Means with different small Latin letter as superscript are not significantly different ($p>0.05$). *Significant distance from the ideal y score.

9.4. Discussion

The starches isolated from sweet potato cv TNG57, TNG66, and TNG73 had higher viscosity, shorter gelatinization time, and lower pasting temperature than the flours from the same cultivars. These results indicated that the gelatinization properties are mainly due to the starch concentration in the samples. The difference in the shape of the curves between the starch samples and those of the flour samples suggested that the pasting behaviors of TNG57, TNG66, and TNG73 may be influenced not only by the starch concentration, but also by some non-starch components. The gelatinization ability of the starches and flours of TNG57, TNG66, and TNG73 enables them to be suitable for development into sweet potato-based new food items, such as cakes, cookies, and puddings.

The sensory evaluation of those products developed from sweet potato flour and starch in this study indicated a good level of overall acceptance, from 5.33 to 7.10 point score in a scale of 0 to 10. With a score ranging from 6.11 to 8.29 in a scale of 0 to 10, the taste was considered as acceptable to the panelists. The panelists indicated that TNG57 and TNG66 spongy cakes had too strong sweet potato flavor and were not sweet enough. These deficiencies are easily correctable to meet the preferences of consumers. The panelists also mentioned that the cookies were too hard and the puddings too fine. Softer cookies can be obtained by adjusting the combination of time and temperature of cooking, while grittier puddings can be prepared by adding flour to the starch. All of these showed that the solid residues of antioxidant extraction from sweet potato can be used to develop new starchy food items. Development of new food items from sweet potato flour and starch can open new opportunities to diversify and increase the food use of sweet potato; thus improving the food and nutrition security in the world, enhancing the revenue of sweet potato farmers, and creating employments via new sweet potato processing factories.

9.5. Conclusions

The starches of TNG57, TNG66, and TNG73 showed high gelatinization ability with peak viscosity higher than 498 RVU occurring before 5 min around 80°C, and the final viscosity higher than 336 RVU. This high gelatinization ability makes TNG57, TNG66, and TNG73 starches suitable for preparing into gel food, as demonstrated by the banana jelly pudding developed from TNG57 starch in this study. Flours of the three cultivars exhibited significantly lower gelatinization ability than the starches. However, they can still be used to develop new food items that require medium gelatinization ability, as demonstrated by the compact cookies and spongy cakes developed from the three cultivars in this study. In a factory, large quantity extraction of antioxidants from sweet potato to be used as food additives requires large volume of sweet potato raw materials, resulting in a large quantity of extraction solid residues. These residues can be exploited to become useful materials for preparing other food items, such as cakes, cookies, and puddings.

Chapter 10. General summary and conclusions

When freshly harvested, TNG57, TNG66, and TNG73 sweet potato cultivars contain approximately 73% moisture. The Kjeldahl crude protein content varied largely, with 2.45% of dry matter in TNG73, 4.49% in TNG57, and 8.96% in TNG66. The three cultivars were not significantly different in term of contents of fiber and calories. TNG57 is a yellow flesh and skin sweet potato. Its yellowness decreased significantly during the drying process. TNG66 has a reddish yellow (orange) flesh and a yellowish red skin. Its flesh color changed into yellow during freeze-drying, LTD, and HAD at 50°C. For TNG73, the flesh was red, and the skin was yellowish red in fresh samples. During HAD at 50ºC, the color became reddish yellow. On average, TNG66 had the highest carotenoids and beta-carotene contents, followed by TNG57, and then TNG73. Due to its high contents of proteins and beta-carotene, TNG66 is a suitable cultivar for mass production as a staple food in developing countries, particularly in areas with vtamin A deficiencies. Introduction of this cultivar to developing countries, where animal protein is expensive and hardly accessible to people with low income, could be worth attempting.

Ethanol and DE extracts exhibited the highest total phenolic contents, followed in descending order by acetone, methanol, EA, aqueous ethanol (1:1 v/v) and ddH_2O extracts. The highest AOA, as evaluated by DPPH assay, were observed in methanol, ethanol, and acetone extracts, followed in descending order by DE, aqueous ethanol, and EA and ddH_2O extracts. However, due to its high effectiveness in etracting anitioxidant from sweet potato, and its relative lower toxicity compared to acetone, DE and methanol, ethanol was chosen for further extractions.

On average, the skin samples had higher phenolic content and AOA than the flesh. As in many cuisines, sweet potato is peeled before cooking, and the skin is considered as a by-product with little value, sweet potato skin can constitute a non-expensive source of antioxidants for food industry to satisfy the general desire of the consumers to replace synthetic food additives with natural alternatives. According to the regression parameters, the phenolic compounds of the fresh samples were responsible for the AOA to the extents of 7.5, 25.8, and 51.2% in TNG57, TNG66, and TNG73, respectively.

TNG73 had, on average, the highest phenolic content and AOA, followed by TNG66, and then TNG57. Freeze-dried samples exhibited higher phenolic content and AOA than the air-dried samples. Similarly to the yellowness of TNG57 and redness of TNG66 and TNG73, the AOA, phenolics, and carotenoid content of the three cultivars decreased significantly during the air-drying process. Higher losses of AOA of TNG57 and TNG66 were observed in hot air drying than low temperature drying. For TNG73 samples, high temperature combined with short time drying resulted in lower losses of AOA in comparison to lower temperature combined to extended drying time.

Three major phenolic antioxidants, namely, catechin, rutin, and gallic acid, were identified in TNG73 ethanol extracts. LLP was a better method to extract gallic acid in comparison to PHM, whereas the latter was better to extract catechin and rutin. Catechin was encountered in the highest concentration, with a maximum content extracted of 268.24 mg/kg dry matter, followed by rutin (67.06 mg/kg dry matter), and then gallic acid (25.56 mg/kg dry matter). Rutin was the major contributor to the AOA of TNG73, followed by catechin, and then gallic acid.

During the storage test of the flour of TNG73, the decrease of the total phenolics of the flesh samples was significant after nine-month storage at room temperature and 10-month storage at 4°C. For the skin samples, the decrease of total phenolics was significant after 11-month storage both at room temperature and 4°C. The decrease of the AOA of the flesh samples was significant after five-month storage at room temperature and 10-moonth storage at 4°C. For the skin samples, the decrease of AOA was significant after 11-month storage both at room temperature and 4°C. The difference between the skin and the flesh samples in maintaining the AOA during storage was attributed to the higher content of phenolics in the skin in comparison to the flesh, as high correlation was observed between the total phenolics and the DPPH IC_{50}. The results suggested that we can process TNG73 into flour to reduce distribution and storage costs and for food security purposes, as the storage can be exploited to provide food availability between at least two consecutive harvests. The results also suggested that for better retention of the phenolic content and AOA, which are important for disease prevention and health promotion by appropriate diet, TNG73 root would better be processed with the skin than without.

The pasting profile analysis indicated that the starches isolated from TNG57, TNG66, and TNG73 had high viscosity, and can be used for preparation of food items that requires moderate to high gelatinization property of the raw material. The flours had lower gelatinization ability than the starches, but they still can be used as raw materials to develop new food items, as suggested by some sweet potato cakes, cookies, and puddings developed in this study. The results showed a good opportunity of using the extraction solid residue of TNG57, TNG66, and TNG73 in the potential case of a factory for natural antioxidant production.

References

Afanas, E. V., A. I. Dorozhka, A. V. Brodskii, V. A. Kostyuk, and A. I. Potapovitch (1989) Chelating and free radical scavenging mechanisms of inhibitory action of rutin and quercetin in lipid peroxidation. Biochem. Pharmacol. 38(11):1763-1769.

Alderman H., J. Behrman, and J. Hoddinott (2004) Improving hild Nutrition for Sustainable Poverty Reduction in Africa. International Food Policy Research Institute (IFPRI), Washington, D.C. 6 pp.

Ansari, N. M., L. Houlihan, B. Hussain, and A. Pieroni (2005) Antioxidant activity of five vegetables traditionally consumed by South-Asian migrants in Bradford, Yorkshire, UK. Phytother. Res. 19: 907–911.

AOAC (1970) Official methods of analysis of the association of analytical chemists (11th edn.). AOAC, Washington, D.C., USA. pp. 1015 pp.

AOAC (2000) Official methods of analysis of the association of analytical chemists (17th edn.). Association of Analytical Communities, Gaithersburg, MD, USA.

Arima, H. K. and D. B. Rodriguez-Amaya (1988) Carotenoid composition and vitamin A value of commercial Brazilian squashes and pumpkins. J. Micronutr. Anal. 4: 177-191.

Arima, H. K. and D. B. Rodriguez-Amaya (1990) Carotenoid composition and vitamin A value of a squash and a pumpkin from Northeastern Brasil. Arch. Latinoam. Nutr. 40: 284-292.

Arnao, M. B., A. Cano, and M. Acosta (1999) Methods to measure the antioxidant activity in plant material. A comparative discussion. Free Rad. Res. 31: 89-96.

Arora, A., M. G. Nair, and G. M. Strasburg (1998) Structure-activity relationships for antioxidant activity of a series of flavonoids in liposomal system. Free Radical Biol. Med. 24 (9):1355-1363.

Aruoma O. I. (1993) Free radicals: their role in nutrition. pp. 187-217. In: Aruoma, O. I. (ed.) Free radicals in tropical diseases. Harwood Academic Publishers, London.

Aruoma, O. I. (1994) Nutrition and health aspects of free radicals and antioxidants. Food Chem. Toxicol. 32: 671-683.

Aruoma, O. I. (1998) Free radicals, oxidative stress, and antioxidants in human health and disease. JAOCS. 75: 199-212.

Aruoma O. I. and B. Halliwell (1991) Free radicals and food additives. Taylor and Francis Ltd., London. 119 pp.

Asami, D. K., Y. J. Hong, D. M. Barrett, and A. E. A. E. Mitchell (2003) Comparison of total phenolics and ascorbic acid content of freeze-dried and air-dried marionberry, strawberry, and corn grown using conventional, organic, and sustainable agricultural practices. J. Agric. Food Chem. 51: 1237-1241.

Auerbach, H. R. and D. A. Gray (1999) Oat antioxidant extraction and measurement. Towards a commercial process. J. Sci. Food Agric. 79: 385-389.

Awad, M. A., A. de Jager, L. H. W. van der Plas, and A. R. van der Krol (2001) Flavonoid and chlorogenic acid changes in peel of 'Elstar' and 'Jonagold' apples during development and ripening. Scientia Hort. 90:69-83.

Bachrach, U. and Y-C Wang (2002) Cancer therapy and prevention by green tea: role of ornithine decarboxylase. Amino Acids. 22: 1-13.

Bartosz, G., A. Janaszewska, D. Ertel, and M. Bartosz (1998) Simple Determination of peroxyl radical-trapping capacity. Biochem. Mol. Biol. Int. 46 (3): 519-528.

BeMiller, J. N. and R. L. Whistler (1996) Carbohydrates. pp. 157-223. In: Fennema O. R. (ed.). Food Chemistry (3rd edn.).

Bendich A. (1990). Carotenoids and the immune system. pp. 323-335. In: Krinsky, N.I., M. M. Mathews-Roth, and R. F. (eds). Carotenoids: Chemistry and Biology. Taylor Plenum Press, NY, USA.

Bendich, A. and J. A. Olson (1989) Biological actions of carotenoids. FASEB J. 3:1927-1932.

Benson, T. (2004) Assessing Africa's Food and Nutrition Security Situation. IFPRI, Washington, D.C. 6 pp.

Benson, T. (2006) Assessing Africa's Food and Nutrition Security Situation. IFPRI, Washington, D.C. 4 pp.

Blasco, A. J., M. C. Rogerio, M. C. Gonzalez, and A. Escarpa (2005) Electrochemical index as a screening method to determine total polyphenolics in foods: A proposal. Anal. Chim. Acta. 539: 237–244.

Block, G., B. Paterson, and A. Subar (1992) Fruit, vegetables and cancer prevention: a review of the epidemiological evidence. Nutr. Cancer. 18: 1-29.

Bloem, M. W., S. de Pee, and I. Darnton-Hill (1998) New issues in developing effective approaches for the prevention and control of vitamin A deficiency. Food Nutr. Bull. 19: 137 – 148.

Blois, M. S. (1958) Antioxidant determinations by the use of a stable free radical. Nature. 181: 1199- 1200.

Bondet, V., W. Brand-Williams, and C. Berset (1997) Kinetics and mechanisms of antioxidant activity using the DPPH free radical method. Food Sci. Technol. 30: 609-615.

Bonorden, W. R. and M. W. Pariza (1994) Antioxidant nutrients and protection from free radicals. pp. 1948. In: Katsonis, F. N., M. Mackey, and J. J. Hjelle (eds.) Nutritional Toxicology. Raven Press, NY, USA.

Borowska, E. J., A. Szajdek, and J. Borowski (2005) Antioxidant properties of fruits, vegetables and their products. Fruit Proces. 15(1): 38-43.

Bouayed, J., K. Piri, H. Rammal, A. Dicko, F. Desor, C. Younos, and R. Soulimani (2007) Comparative evaluation of the antioxidant potential of some Iranian medicinal plants. Food Chem. 104: 364–368.

Bovell-Benjamin, A. C. (2007) Sweet potato: a review of its past, present, and future role in human nutrition. Adv. Food Nutr. Res. 52:1-59.

Brand-Williams, W., M. E. Cuvelier, and C. Berset (1995) Use of a free radical method to evaluate antioxidant activity. Food Sci. Technol. 28: 25-30.

Britton, G. (1991) Carotenoids. Methods Plant Biochem. 7:473-518.

Brownridge, E. (1993) Food that fights back: antioxidants are nutrition's newest superheroes. Pharm. Prac. 9:40-50.

Burns, J., P. T. Gardner, D. Matthews, G. G. Duthie, M. E. Lean, and A. Crozier (2001) Extraction of phenolics and changes in antioxidant activity of red wines during vinification. J. Agric. Food Chem. 49: 5797-5808.

Burton, G. W. (1989) Antioxidant action of carotenoids. J. Nutr. 119:109-111.

Burton, G. W. and K. U. Ingold (1984) β-carotene: An unusual type of lipid antioxidant. Science. 224: 569-573.

Byers, T. and G. Perry (1992) Dietary carotenes, vitamin C, and vitamin E as protective antioxidants in human cancers. Ann. Rev. Nutr. 12:139-159.

Cano, A., J. HernKndez-Rulz, E García-Cánovas, M. Acosta and M.B. Arnao (1998) An end-point method for estimation of the total antioxidant activity in plant material. Phytochem. Anal. 9:196-202.

Cano, A., O. Alcaraz, and M. B. Arnao (2003) Free radical-scavenging activity of indolic compounds in aqueous and ethanolic media. Anal. Bioanal. Chem. 376: 33–37.

Cao, G., C.P. Verdon, A.H.B. Wu, H. Wang, and R.L. Prior (1995) Automated assay of oxygen radical absorbance capacity with the Cobas Fara II. Clin. Chem. 41: 1738-1744.

Cao, G., H. M. Alessio, and R.G. Cutler (1993). Oxygenradical absorbance capacity assay for antioxidants. Free Radic. Biol. Med. 14 (3): 303-311.

CGIAR. (2006) Research and Impact: area of research: sweet potato. Retrieved February 6, 2007, from the World Wide Web: [http://www. cgiar.org/ impact/ research/ sweet potato. html.

Chandler, L. A. S. J. Schwartz (1998) Isomerization and losses of trans-carotene in sweet potatoes as affected by processing treatments. J. Agric. Food Chem. 36: 129–133.

CIELab (2003) Ets wine color analysis. (Ets laboratories, 15 October 2003). Retrieved January 1, 2006, from the World Wide Web: http://www. etslabs. com/scripts/ets/ ContentDocs/

CIP (1998) Publications / Annual Report 1998. Diversifying Diets in China. Retrieved March 10, 2007, from the World Wide Web: http:// www. cipotato.org/sweetpotato/.

Cooper, T. G. (1977) The tools of biochemistry (1st edn). John Wiley and Sons, NY, USA. 423 pp.

Council of Agriculture, Executive Yuan, R.O.C. (2006) Sweet potato. Retrieved February 28, 2007, from the world wide web: http://eng.coa.gov.tw/

Craft, N. E. and J. H. Jr. Soares (1992) Relative solubility, stability, and absorptivity of lutein and β-carotene in organic solvents. J. Agric. Food Chem. 40:431-43.

Damodaran, S. (1996) Amino-acids, peptides, and proteins. pp. 321-429. In: Fennema O. R. (ed). Food Chemistry (3rd edn.).

Davies, B. H. (1976) Carotenoids. pp 38-165. In: Goodwin TW (ed). Chemistry and biochemistry of plant pigments (2nd edn.). Academic Press, London.

de Arantes-Frota A. C., A. A. F. de Souza-Lima, A. B. J. Dias, A. C. dos Santos-Lourenço, E. Antecka, and M. N. J. Burnier (2008) Freeze-Drying as an Alternative Method of Human Sclera Preservation. Arquivos Brasileiros de Oftalmologia. 71(2): 137-141.

Diaz, M. N., B. Frei, J. A. Vita, and J. F. Keaney (1997) Antioxidants and atherosclerotic heart disease. N. Engl. J. Med. 337: 408-416.

Dietrich, H., A. Rechner, and C. D. Patz (2004) Bioactive compounds in fruit and juice. Fruit Proces. 14: 50-55.

Di Mascio, P., S. Kaiser, and H. Sies (1989) Lycopene as the most efficient biological carotenoid singlet oxygen quencher. Arch. Biochem. Biophys. 274:532-538.

Di Venere, D., L. V. Sergio, and V. V. Bianco (2005) Antioxidant phenolics in escarole and radicchio during storage of fresh-cut 'ready-to-use' Product. Acta Hort. 682: 1947-1951.

Duvivier P., Hsieh P.-C., Lai P.-Y., and Charles A. L. (2008a) Total phenolic content and antioxidant activity of seven extracts of sweet potato (*Ipomoea batatas* Lam.) as affected by the extracting solvent and the genotype. Recherche, Etude, dévelopement (RED) 4 (1): 20-27.

Duvivier P., Hsieh P.-C., Lai P.-Y., and Charles A. L. (2008b). Evaluation of drying methods on antioxidant activity, total phenolic and total carotenoid contents of sweet potato (*Ipomoea batatas* Lam.) var. Tainong 73. J. International Cooperation (JIC) 3(1): 73-86.

Duvivier P., P.-C. Hsieh, P.-Y. Lai, and A. L. Charles (2009) Effect of freeze-drying on phenolic content and antioxidant activity of Tainong 57, Tainong 66, and Tainong 73 cultivars of sweet potato (*Ipomoea batatas* Lam.). J. International Cooperation (JIC) 4(1): 1-15.

Ehlenfeldt, M. K. and R. L. Prior (2001) Oxygen radical absorbance capacity (ORAC) and phenolic and anthocyanin concentrations in fruit and leaf tissues of high-bush blueberry. J. Agric. Food Chem. 49:2222-2227.

Ekanayake, P. M., G. T. Park, Y. D. Lee, S. J. Kim, S. C. Jeong, and J. Lee (2005) Antioxidant potential of eel (Anguilla Japonica and Conger Myriaster) flesh and skin. J. Food Lipids 12: 34–47.

Elbe, J. H. V., and S. J. Schwartz (1996) Colorants. pp. 651-722. In: Fenema O. R. (ed.). Food Chemistry (3rd edn.).

Etkin, N. L. (1996) Medicinal cuisines: diet and ethnopharmacology. Int. J. Pharmacol. 34: 313–326.

European Parliament (2002) Regulation (EC) No 178/2002 of European Parliament (28 January 2002). 37 pp.
FAO (1984) Production yearbook 1983. FAO, Rome. Retrieved December 2, 2006, from the World Wide Web: http://www.fao.org/ newsroom/ en/ focus.
FAO (2003) Trade reforms and food security. FAO, Rome. 296 pp.
FAO (2004) Understanding food insecurity. The state of food insecurity in the world. FAO, Rome. 43 pp. Retrieved December 2, 2006, from the World Wide Web: http://www.fao.org/ newsroom/en/focus/ 2004/ 51786/ index. html.
FAO (2006a) The state of food insecurity in the world 2005. FAO, Rome. 9 pp. Retrieved October 22, 2006, from the World Wide Web: http://en.wikipedia.org/ wiki/ Food.
FAO (2006b) FAO Contrystat. Repiblik d Ayiti. Retrieved October 21, 2006, from the World Wide Web: http://www.fao.org/ countryProfiles/ default. asp?lang=en
FAO and WHO (2002) Joint WHO/FAO expert consultation on diet, nutrition and the prevention of chronic diseases. WHO technical report series 916.
Geneva, 28 Jan.-1 Feb. 2002. Retrieved on Oct. 21, 2006, from the World Wide Web: http://www.who.int/nutrition/topics/ dietnutrition_ and_chronicdiseases/en/print.html
FAOSTAT (2005). Food and Agricultural Statistical Database. Retrieved January 18, 2008, from the World Wide Web: http://faostat.fao.org.
FAO / WHO / UNU (1985) Energy and protein requirements. WHO Technical Report Series 724. Geneva: WHO. 301 pp.
FDA (2006) Federal Food, Drug, and cosmetic act. Retrieved October 21, 2006, from the World Wide Web: http://www.fda. gov/ opacom/ laws/ fdcact/ fdctoc.htm
Fennema, O. R. (1996) Water and Ice. pp. 17-94. In: Fennema O. R. (ed). Food Chemistry (3rd edn.).
Fogliano, V., V. Verde, G. Randazzo, and A. Ritieni (1999) Method for measuring antioxidant activity and its application to monitoring the antioxidant capacity of wines. J. Agric. Food Chem. 1999, 47, 1035-1040.
Folin, O. and V. Ciocalteu (1927) On tyrosine and tryptophane determination in proteins. J. Biol. Chem. 72 (2): 627–650.
Foote, C. S., Y. C. Chang, and R. W. Denny (1970) Chemistry of singlet oxygen X. Carotenoid quenching parallels biological protection. J. Am. Oil Chem. Soc. 92:5216-5218.
Garret, E. H. (2002) Fresh-cut Produce: Tracks and Trends. pp. 1-10. In: in Lamikanra, O. (ed.). Fresh Cut Fruits and Vegetables, Science, Technology, and Market. Florida, USA: CRC Press.
Gerster, H. (1991) Potential role of beta-carotene in the prevention of cardiovascular disease. Int. J. Vit. Nutr. Res. 61:277-291.
Goh, S.H. (1996) Carotene controversy. Malaysian Oil Sci. Technol. 5: 17.

Gomes, K. S. M. (2007) Extração e uso de corantes vagetais da amâzonia no tingimendo do couro da Matrinxiã (Brycon amazonicu, Spix & Agassiz, 1819). Master thesis, Instituto Nacional de Pesquisa da Amâzonia, Universidade Federal da Amâzonias, Brazil.

Goodman, G., E. Thornquist, J. Balmes, M. R. Cullen, F. L. Meyskens Jr, G. S. Omenn, B. Valanis, and J. H. Williams Jr. (2004) The β-carotene and retinol efficacy trial: incidence of lung cancer and cardiovascular disease mortality during 6-year follow-up after stopping β-carotene and retinol supplements. J. Natl. Cancer Inst. 96:1743-1750.

Grabowski, S., M. Marcotte, M. Poirier, and T. Kudra (2002) Drying characteristics of osmotically pretreated cranberries – energy and quality aspects. Drying Technology. 20 (10): 1989-2004.

Gregory III, J. F. (1996) Vitamins. pp. 531-616. In: Fennema O. R. (ed). Food Chemistry (3rd edn.).

Grey, K. F. (1990) The antioxidant hypothesis of cardiovascular disease: epidemiology and mechanisms. Biochem. Soc. Trans. 18: 1041-1045.

Gross, J., R. Ikan, and G. Eckhardt (1983) Carotenoids of the fruit of Overrhoa carambola. Phytochem. 22:1479-1481.

Guo, J.-T., H.-L. Lee, S.-H. Chiang, F. I. Lin, and C.-Y. Chang (2001) Antioxidant properties of the extracts from different parts of broccoli in Taiwan. J. Food Drug. Anal. 9(2): 96-101.

Halliwell, B. and J. M. C. Gutteridge (1989) Free radicals in biology and medicine (2nd edn.). Clarendon Press, Oxford. 543 pp.

Heidhues, F., A. Atsain, H. Nyangito, M. Padilla, G. Ghersi, and J.-C. Le Vallée (2004) Development strategies and Africa's food and nutrition security. IFPRI, Washington, D C.60 pp.

Heinäaho, M., J. Pusenius, and R. Julkunen-Tiitto (2006) Organic farming methods affect the concentration of phenolic compounds in sea buckthorn leaves. Joint Organic Congress, Odense, Denmark, May 30-31, 2006.
Retrieved May 21, 2008, from the World Wide Web: http://orgprints.org/7382/

Hertog, M. G. L., E. J. M. Feskens, P. C. H. Hollman, M. B. Katan, and D. Kromhout (1993) Dietary antioxidant flavonoids and risk of coronary heart disease: Lancet. 342: 1007-1011.

Hendren, E., V. Diaz, and U. Svanberg (2002) Estimation of carotenoid accessibility from carrots determined by an in vitro digestion method. Eur. J. Clin. Nutr. 56: 425–430.

Hertog, M. G. L., E. J. M. Feskens, P. C. H. Hollman, M. B. Katan, and D. Kromhout (1993) Dietary antioxidant flavonoids and risk of coronary heart disease: Lancet. 342: 1007-1011.

Higashi-Okai, K., M. Yamazaki, H. Nagamori, and Y. Okai (2001) Identification and antioxidant activity of several pigments from the residual green tea (Camellia sinensis) after hot water extraction. J. UOEH. 23(4):335-44.

Higby, W. K. A. (1962) A simplified method for determination of some aspects of the carotenoids distribution in natural and carotene fortified orange juice. J. Food Sci. 27: 42–49.

Higley, J. S., S. L. Love, W. J. Price, J. E. Nelson, and K. C. Huber (2003) The rapid visco analyzer (RVA) as a tool for differentiating potato cultivars on the basis of flour pasting properties. Amer. J. Potato Res. 80:195-206.

Hogg, J. S., D. H. Lohmann and K. E. Russell (1961) The kinetics of reaction of 2,2-diphenyl-l-picrylhydrazyl with phenols. Can. J. Chem. 39:1588-1594.

Houghton, P. J. and A. Raman (1998) Laboratory handbook for the fractionation of natural extracts (1st edn). Chapman and Hall, NY, USA. 199 pp.

Howe, P. and S. Devereux (2004) Famine intensity and magnitude scales: A proposal for an instrumental definition of famine. Disasters, 28 (4): 353-372.

Huaman, Z. and D. P. Zhang (1997) Sweet potato. pp. 29-38. In: D. Fuccillo *et al.* (Eds). Biodiversity in trust – Conservation and use of plant genetic resources in CGIAR centers. Cambridge University Press, Cambridge, UK.

Huang, D.-J., C.-D. Lin, H.-J. Chen, and Y.-H. Lin (2004) Antioxidant and antiproliferative activities of sweet potato (*Ipomea batatas* [L.] Lam.) 'Tainong 57' contituents. Bot. Bull. Acad. Sinica. 45:179-186.

IFPRI (1995) A 2020 vision for food, agriculture, and the environment: The vision, challenge, and recommended action. IFPRI, Washington, D.C. 23 pp.

Islam, M. S., M. Yoshimoto, S. Yahara, S. Okuno, K. Ishiguro, and O. Yamakawa (2002) Identification and characterization of foliar polyphenolic composition in sweet potato (*Ipomoea batatas* L.) genotypes. J. Agric. Food Chem. 50: 3718-3722.

Jenner, P. (1991) Oxidative stress as a cause of Parkinson's disease. Acta Neurol. Sand. 84(136): 6-15.

Jung, C. H., H. M. Seog, I. W. Choi, and H. Y. Cho (2005) Antioxidant activities of cultivated and wild Korean ginseng leaves. Food Chem. 92: 535-540.

Kazantzis, G. (2004) Cadmium, osteoporosis and calcium metabolism. Biometals 17 (5): 493-498.

Kazutoshi, O. (2004) Germplasm enhancement and breeding strategies for crop quality in Japan. Genebank, National Institute of Agrobiological Sciences, Japan. Retrieved May 21, 2008, from the World Wide Web: http:www/nias.affrc.go.jp, okusan@affrc.go.jp.

Kendall, P. (2000) Good Food Sources of Antioxidants. Colorado State University, Cooperative Extension.

Kessler, G. J. (2000) The bone density diet. Retrieved on Dec. 2, 2006, from the World Wide Web: http://www.ncbi.nlm.nih.gov/ entrez/ query. fcgi? cmd

Khachik, F., G. R. Beecher, and B. G. Mudlagiri (1991) Separation, identification, and quantification of carotenoids in fruits, vegetables, and human plasma by high performance liquid chromatography. Pure Appl. Chem. 63(1): 71-80.

Kim J.-K., J. H. Noh, S. Lee, J. S. Choi, H. Suh, H. Y. Chung, Y.-O. Song, and W. C. Choi (2002) The first total synthesis of 2, 3, 6-tribromo-4,5-dihydroxybenzyl methyl ether (TDB) and its antioxidant activity. Bull. Korean Chem. Soc. 23(5): 661-662.

Kim, K. H., R. Tsao, R. Yang, and S. W. Cui (2005) Phenolic acid profile and antioxidant activity of wheat bran and the effect of hydrolysis conditions. Food Chem. 95: 466–473.

KMA, GTZ, and FAO (1998) Post-harvest systems of potato and sweet potato in Kenya - Final report, Nairobi, Kenya - June, 1998.

Kondo, S., K. Tsuda, N. Muto, and J. Ueda (2002) Antioxidative activity of apple peel or flesh extracts associated with fruit development on selected apple cultivars. Sci. Hort. 96:177-185.

Kooijman L. M., K. J. Ganzeveld, R. M. Manurung, and H. J. Heeres (2003) "Experimental Studies on the Carboxymethylation of Arrowroot Starch in Isopropanol-Water Media." Starch, 55: 495–503.

Kozai, T., C. Kubota, and Y. Kitaya (1996) Sweet potato technology for solving the global issues on food, energy, natural resources, and environmental in 21st century. Environ Control Biol. 34: 105-114.

Krinsky, N. I. (1989) Antioxidant functions of carotenoids. Free Radical Biol. Med. 7:617-635.

Krinsky, N. I. (1990) Carotenoids in medicine. pp. 279-291. In: Krinsky, N.I., M. M. Mathews-Roth, and R. F. Taylor (eds). Carotenoids: Chemistry and biology. Plenum Press, NY, USA.

Krinsky, N. I. (1994) The biological properties of carotenoids. Pure Appl. Chem. 66:1003-1010.

Kuehl, O. R. (1999) Design of experiments: Statistical principles of research design and analysis (2nd edn). Brooks/Cole, Florence, KY, USA. 688 pp.

Kuntz, L.A. (1992) Keeping microorganisms in control. Food Product Design. August: 44-51.

Kurata, R., M. Adachi, O. Yamakawa, and M. Yoshimoto (2007) Growth suppression of human cancer cells by polyphenolics from sweet potato (*Ipomoea batatas* L.) leaves. J. Agric. Food Chem. 55: 185–190.

Ladislav, F., P. Vera, S. Karel, and V. Karel (2005) Reliability of carotenoid analyses: A review. Curr. Anal. Chem. 1: 93-102.

Lana, M. M. (2006) Effects of cutting and maturity on antioxidant activity of fresh–cut tomatoes. Food Chem. 97(2): 203-211.

Lana, M.M. (2005) Modelling quality of fresh-cut tomato based on stage of maturity and storage conditions. Ph.D. Thesis. Wageningen University. With references and summary in English, Dutch and Portuguese. Horticultural production Chains Group, Plant Science, Wageningen University, Wageningen, The Netherlands. 208 pp.

Lappé, F. M. (1976) Diet for a small planet (2nd edn.) Ballantine Books, NY, USA. 411 pp.

Laranjinha, J. A. N., L. M. Almeida, and V. M. C. Madeira (1994) Reactivity of dietary phenolic acids with peroxyl radicals: antioxidant activity upon low density lipoprotein peroxidation. Biochem. Pharmacol. 48: 487-494.

Larraruri, J. A., C. Sanchez-Moreno, and F. Saura-Calixto (1998) Effect of temperature on the free radical scavenging capacity of extracts from red and white grape pomace peels. J. Agric. Food Chem. 46: 2694–2697.

Larsen, E. and L. P. Christensen (2005) Simple saponification method for the quantitative determination of carotenoids in green vegetables. J. Agric. Food Chem. 53 (17): 6598 -6602.

Lebeau, J., C. Furman, J.-L. Bernier, P. Duriez, E. Teissier, and N. Cotelle. (2000) Antioxidant properties of di-tert-butylhydroxylated flavonoids. Free Radic. Biol. Med. 29(9): 900-912.

Leitão, G.G., S. G. Leitao, and W. Vilegas (2002) Quick preparative separation of natural naphthoquinones with antioxidant activity by high-speed counter-current chromatography. Z. Naturforsch. 57: 1051-1055.

Li, P., H. Anu, S. Jari , Y. Teijo, and V. Heikki (2006) TLC method for evaluation of free radical scavenging activity of rapeseed meal by video scanning technology. In: Proceedings of the 10th International Rapeseed Congress, Camberra, Australia.

Lindsay, R. C. (1996) Flavors. pp. 723-765. In: Fennema O. R. (ed). Food Chemistry (3rd edn.).

Lineback, D. R. and C. H. Ke (1975) Starches and low-molecular -weight carbohydrates from chick pea and horse bean flours. Cereal Chemistry, 52:334-347.

Liu, C. L., Y. S. Chen, J. H. Yan, and B. H. Chang (2008) Antioxidant activity of tartary (*Fagopyrum tartaricum* (L) Gaerth) and common (*Fagopyrum esculentum* Moench) buckwheat sprouts. J. Agric. Food Chem. 56:173-178.

Lo, S.-F., S. M. Nalwade, V. Mulabagal, S. Matthew, C.-L. Chen, C. –L. Kuo, and H.-S. Tsay (2004) In vitro propagation by asymbiotic seed germination and 1,1-diphenyl-2-picrylhydrazyl (DPPH) radical scavenging activity 66, and Tainong 73 cultivars of sweet potato (*Ipomoea batatas* Lam.). J. International Cooperation (JIC) 4(1): 1-15.

Ehlenfeldt, M. K. and R. L. Prior (2001) Oxygen radical absorbance capacity (ORAC) and phenolic and anthocyanin concentrations in fruit and leaf tissues of high-bush blueberry. J. Agric. Food Chem. 49:2222-2227.

Ekanayake, P. M., G. T. Park, Y. D. Lee, S. J. Kim, S. C. Jeong, and J. Lee (2005) Antioxidant potential of eel (Anguilla Japonica and Conger Myriaster) flesh and skin. J. Food Lipids 12: 34–47.

Elbe, J. H. V., and S. J. Schwartz (1996) Colorants. pp. 651-722. In: Fenema O. R. (ed.). Food Chemistry (3rd edn.).

Etkin, N. L. (1996) Medicinal cuisines: diet and ethnopharmacology. Int. J. Pharmacol. 34: 313–326.

European Parliament (2002) Regulation (EC) No 178/2002 of European Parliament (28 January 2002). 37 pp.
FAO (1984) Production yearbook 1983. FAO, Rome. Retrieved December 2, 2006, from the World Wide Web: http://www.fao.org/ newsroom/ en/ focus.
FAO (2003) Trade reforms and food security. FAO, Rome. 296 pp.
FAO (2004) Understanding food insecurity. The state of food insecurity in the world. FAO, Rome. 43 pp. Retrieved December 2, 2006, from the World Wide Web: http://www.fao.org/ newsroom/en/focus/ 2004/ 51786/ index. html.
FAO (2006a) The state of food insecurity in the world 2005. FAO, Rome. 9 pp. Retrieved October 22, 2006, from the World Wide Web: http://en.wikipedia.org/ wiki/ Food.
FAO (2006b) FAO Contrystat. Repiblik d Ayiti. Retrieved October 21, 2006, from the World Wide Web: http://www.fao.org/ countryProfiles/ default. asp?lang=en
FAO and WHO (2002) Joint WHO/FAO expert consultation on diet, nutrition and the prevention of chronic diseases. WHO technical report series 916. Geneva, 28 Jan.-1 Feb. 2002. Retrieved on Oct. 21, 2006, from the World Wide Web: http://www.who.int/nutrition/topics/ dietnutrition_ and_chronicdiseases/en/print.html
FAOSTAT (2005). Food and Agricultural Statistical Database. Retrieved January 18, 2008, from the World Wide Web: http://faostat.fao.org.
FAO / WHO / UNU (1985) Energy and protein requirements. WHO Technical Report Series 724. Geneva: WHO. 301 pp.
FDA (2006) Federal Food, Drug, and cosmetic act. Retrieved October 21, 2006, from the World Wide Web: http://www.fda. gov/ opacom/ laws/ fdcact/ fdctoc.htm
Fennema, O. R. (1996) Water and Ice. pp. 17-94. In: Fennema O. R. (ed). Food Chemistry (3rd edn.).
Fogliano, V., V. Verde, G. Randazzo, and A. Ritieni (1999) Method for measuring antioxidant activity and its application to monitoring the antioxidant capacity of wines. J. Agric. Food Chem. 1999, 47, 1035-1040.
Folin, O. and V. Ciocalteu (1927) On tyrosine and tryptophane determination in proteins. J. Biol. Chem. 72 (2): 627–650.
Foote, C. S., Y. C. Chang, and R. W. Denny (1970) Chemistry of singlet oxygen X. Carotenoid quenching parallels biological protection. J. Am. Oil Chem. Soc. 92:5216-5218.
Garret, E. H. (2002) Fresh-cut Produce: Tracks and Trends. pp. 1-10. In: in Lamikanra, O. (ed.). Fresh Cut Fruits and Vegetables, Science, Technology, and Market. Florida, USA: CRC Press.
Gerster, H. (1991) Potential role of beta-carotene in the prevention of cardiovascular disease. Int. J. Vit. Nutr. Res. 61:277-291.
Goh, S.H. (1996) Carotene controversy. Malaysian Oil Sci. Technol. 5: 17.

Gomes, K. S. M. (2007) Extração e uso de corantes vagetais da amâzonia no tingimendo do couro da Matrinxiã (Brycon amazonicu, Spix & Agassiz, 1819). Master thesis, Instituto Nacional de Pesquisa da Amâzonia, Universidade Federal da Amâzonias, Brazil.

Goodman, G., E. Thornquist, J. Balmes, M. R. Cullen, F. L. Meyskens Jr, G. S. Omenn, B. Valanis, and J. H. Williams Jr. (2004) The β-carotene and retinol efficacy trial: incidence of lung cancer and cardiovascular disease mortality during 6-year follow-up after stopping β-carotene and retinol supplements. J. Natl. Cancer Inst. 96:1743-1750.

Grabowski, S., M. Marcotte, M. Poirier, and T. Kudra (2002) Drying characteristics of osmotically pretreated cranberries – energy and quality aspects. Drying Technology. 20 (10): 1989-2004.

Gregory III, J. F. (1996) Vitamins. pp. 531-616. In: Fennema O. R. (ed). Food Chemistry (3rd edn.).

Grey, K. F. (1990) The antioxidant hypothesis of cardiovascular disease: epidemiology and mechanisms. Biochem. Soc. Trans. 18: 1041-1045.

Gross, J., R. Ikan, and G. Eckhardt (1983) Carotenoids of the fruit of Overrhoa carambola. Phytochem. 22:1479-1481.

Guo, J.-T., H.-L. Lee, S.-H. Chiang, F. I. Lin, and C.-Y. Chang (2001) Antioxidant properties of the extracts from different parts of broccoli in Taiwan. J. Food Drug. Anal. 9(2): 96-101.

Halliwell, B. and J. M. C. Gutteridge (1989) Free radicals in biology and medicine (2nd edn.). Clarendon Press, Oxford. 543 pp.

Heidhues, F., A. Atsain, H. Nyangito, M. Padilla, G. Ghersi, and J.-C. Le Vallée (2004) Development strategies and Africa's food and nutrition security. IFPRI, Washington, D C.60 pp.

Heinäaho, M., J. Pusenius, and R. Julkunen-Tiitto (2006) Organic farming methods affect the concentration of phenolic compounds in sea buckthorn leaves. Joint Organic Congress, Odense, Denmark, May 30-31, 2006.

studies of tissue culture raised plants of three medicinally important species of Dendrobium. Biol. Pharm. Bull. 27(5): 731-735.

Loliger, J. 1991. The use of antioxidants in foods. pp. 121-1 50. In: Aruoma O. I. and B. Halliwell (eds). Free Radicals and Food Additive. Taylor and Francis, London.

Lowry, O H, N. J. Rosebrough, A. L. Farr, and R. J. Randall (1951) Protein measurement with the Folin phenol reagent. J. Biol. Chem. 193: 265-275.

Lu, Y. and L. Y. Foo (2000) Antioxidant and radical scavenging activities of polyphenols from apple pomace. Food Chem. 68: 81-85.

Marc, F., A. Davin, L. Deglène-Benbrahim, C. Ferrand, M.Baccaunaud, P. Fritsch (2004) Méthodes d'évaluation du potentiel antioxydant dans les aliments. Med. Sci. 20: 458-63.

Marinova, D., F. Ribarova, and M. Atanassova (2005) Total phenolics and total flavonoids in Bulgarian fruits and vegetables. J. U. Chem. Technol Metall. 40 (3): 255-260.

Mathews-Roth, M. M. (1985) Carotenoid and cancer prevention — experimental and epidemiological studies. Pure Appl. Chem. 57: 717-722.

Mathews-Roth, M. M. (1991) Recent progress in the medical applications of carotenoids. Pure Appl. Chem. 63: 147-156.

McGee, H. J. (2004) On food and cooking. Scribner. ISBN 0684800012. 896 pp. Retrieved October 21, 2006, from: http://en. wikipedia.org/wiki/ Food#Legal_definition

Meloan, C. E and Y. Pomeranz (1980) Food analysis laboratory experiments, (2nd edn.). The AVI publishing company, Inc. Westport, Connecticut. 157 pp.

Mercadante, A. Z., D. B., Rodriguez-Amaya, and G. Britton (1997) HPLC and mass spectrometric analysis of carotenoids from mango. J. Agric. Food Chem. 45:120-123.

Mercadante, A. Z., G. Britton, and D.B. Rodriguez-Amaya (1998) Carotenoids from yellow passion fruit (*Passiflora edulis*). J. Agric. Food Chem. 46:4102-4106.

Miliauskas, G., P. R. Venskutonis, and T. A. van Beek (2004) Screening of radical scavenging activity of some medicinal and aromatic plant extracts. Food Chem. 85: 231–237.

Miller, D. D. (1996) Minerals. pp. 617-649. In: Fennema O. R. (ed). Food Chemistry (3rd edn.).

Miller, N. J. and C.A. Rice-Evans (1997) Factors influencing the antioxidant activity determined by the ABTS radical cation assay. Free radical Res. 26:195-199.

Miller, N. J., C. Rice-Evans, M. J. Davies, V. Gopinathan, and A. Milner (1993) A novel method for measuring antioxidant capacity and its application to monitoring the antioxidant status in premature neonates. Clin. Sci. 84: 407-412.

Millward, D. J., Newsholme, E. A. Pellett, P. L. and R. Uauy (1992). Amino acid scoring in health and disease. In: Protein-Energy Interactions - Proceedings of a workshop held by the International Dietary Energy Consultancy Group. Switzerland: IDECG.

Molyneux, P. (2004) The use of the stable free radical diphenylpicrylhydrazyl (DPPH) for estimating antioxidant activity. Songklanakarin J. Sci. Technol. 26(2) : 211-219.

Moure, A., J. M. Cruz, D. Fanco, M. Dominguez, J. Sineiro, H. Dominguez, J. Nunez, and J. C. Parajo (2001) Natural antioxidants from residual sources. Food Chem. 72: 145–171.

Nawar, W. W. (1996) Lipids. pp. 225-319. In: Fennema O. R. (ed). Food Chemistry (3rd edn.).

Muchoki C. N., J. K. Imungi, and P. O. Lamuka (2007) Changes in beta-carotene, ascorbic acid and sensory properties in fermented, solar-dried and stored cowpea leaf vegetables. African Journal of Food, Agriculture, Nutrition and Development (AJFAND) 7(3). Online journal. Accessible at www.ajfand.net

Nagao, T., Y. Komine, S. Soga, S. Meguro, T, Hase, Y. Tanaka, and I. Tokimitsu (2005) Ingestion of a tea rich in catechins leads to a reduction in body fat and malondialdehyde-modified LDL in men. Am. J. Clin. Nutr. 81:122–129.

NRLO. (2001) Functional Foods: Position and future perspectives, report no 2000/15. The Hague, February 2001. ISBN: 90 - 5059 - 121 - 3. 79 pp.

O'Brien, P.J. (1972) The sweet potato: its origin and dispersal. Am. Anthropologist. 74: 343-365.

Ooyanagui, Y. (1984) Re-evaluation of assay methods and established of kit for superoxide dismutase activity. Anal. Biochem. 142:290-296.

Padda M. S. (2006) Phenolic Composition and Antioxidant Activity of Sweetpotatoes [*Ipomoea batatas* (L.) Lam.]. PhD thesis. Louisiana USA: Louisiana State University and Agricultural and Mechanical College.

Palozza, P, and N. I. Krinsky (1992) Antioxidant effects of carotenoids in vivo and in vitro: An overview. Methods Enzymol. 213:403-420.

Pardey, P. G., J. M. Alston, and R. R. Piggott (2006) Agricultural R&D in the developing world: Too little, too late? International Food Policy Research Institute (IFPRI), Washington, D.C.Washington, DC, USA. 22 pp.

Pariza, M. W. (1996) Toxic substances. pp. 825-840. In: Fennema O. R. (ed). Food Chemistry (3rd edn.).

Picha, D. H. (1985) Crude protein, minerals, and total carotenoids in sweet potatoes. J. Food Sci. 50: 1768-1769.

Pieroni, A. (2000) Medicinal plants and food medicines in the folk traditions of the upper Lucca Province, Italy. J Ethnopharmacol. 70: 235–273.

Pieroni, A., and L. L. Price (2005) Eating and healing: Traditional food as medicine. Haworth Press, Binghamton, NY, USA. 480 pp.

Pieroni, A., C. Nebel, C. Quave, H. Münz, and M. Heinrich. (2002). Ethnopharmacology of liakra: traditional weedy vegetables of the Arbëreshë of the Vulture area in southern Italy. J Ethnopharmacol. 81: 165–185.

Pieroni, A., V. Janiak, C. M. Dürr, S. Lüdeke, E. Trachsel, and M. Heinrich (2002). In vitro antioxidant activity of non-cultivated vegetables of ethnic Albanians in southern Italy. Phytother. Res. 16: 467–473.

Pinstrup-Andersen, P. (2001) The future world food situation and the role of plant diseases. The Plant Health Instructor. [online], 10 (1094). Retrieved October 28, 2006, from: http://en. wikipedia. org/ wiki /Food.

Potter, N. N. (1986) Food science (4th edn.). Avi publishing company, Inc. Westport Connecticut. 735 pp.

Prince, R. L., A. Devine, and S. S. Dhaliwal (2006) Effects of calcium supplementation on clinical fractures and bone structure: Results of a 5-year, double-blind, placebo-controlled trial in elderly women. Arch. Intern. Med. 166: 869-875.

Prior, R. L., G. Cao, A. Martin, E. Sofic, J. McEwen, C. O'Brien, N. Lischner, M. Ehlenfeldt, W. Kalt, G. Krewer, and C. M. Mainland (1998)

Antioxidant capacity as influenced by total phenolic and anthocyanin content, maturity, and cultivar of Vaccinium species. J. Agric. Food Chem. 46: 2686-2693.

Puah, C. W., Y. M. Choo, N. A. Ma, and C. H. Chuah (2005) Supercritical fluid extraction of palm carotenoids. Am. J. Environ. Sci. 1 (4): 264-269.

Quackenbush, F. W., and R. L. Smallidge (1986) Nonaqueous reverse phase liquid chromatographic system for separation and quantitation of provitamins A. J. Assoc. Off. Anal. Chem. 69:767-772.

Raina, K., S. Rajamanickam, G. Deep, M. Singh, R. Agarwal, and C. Argawal (2008) Chemopreventive effects of oral gallic acid feeding on tumor growth and progression in TRAMP mice. Mol. Cancer Ther. 7(5):1258–1267.

Raisz, L. (2005) Pathogenesis of osteoporosis: concepts, conflicts, and prospects. J. Clin. Invest. 115 (12): 3318-3325.

Randhir, R. and K. Shetty (2005). Developmental stimulation of total phenolics and related antioxidant activity in light and dark-germinated corn by natural elicitors. Process Biochem. 40: 1721−1732.

Ratti, C. (2001) Hot air and freeze-drying of high-value foods: a review. J. Food Eng. 49: 311-319.

Ravindran, V., G. Ravindran, R. Sivakanesan, and S. B. Rajaguru (1995) Biochemical and nutritional assessment of tubers from 16 cultivars of sweetpotato (*Ipomoea batatas* L.). J. Agric. Food Chem. 43: 2646-2651.

Rehman, Z., A. M. Salariya, and F. Habib (2003) Antioxidant activity of ginger extract in sunflower oil. J. Sci. Food Agric. 83: 624–629.

Rembiałkowska, E. (2006) Organic food quality – axioms and ambiguities. Paper presented at Joint Organic Congress, Odense, Denmark, May 30-31, 2006.

Ribeiro, S. M. R., J. H. de Queiroz, M. E. L. R. de Queiroz, F. M. Campos, and H. M. S. Pinheiro (2007) Antioxidant in mango (Mangifera indica L.) pulp. Plant Foods Hum. Nutr. 62: 13-17.

Rice-Evans, C., N. J. Miller, and G. Paganga (1995) Structure-antioxidant activity relationships of flavonoids and phenolic acids. Free Radical Biol. Med. 20 (7): 933-956.

Riemersma, R. A., D. A. Wood, C. E. MacIntyre, R. A. Elton, K. F. Grey, and M. F. Oliver (1991) Risk of angina pectoris and plasma concentrations of vitamin A, C and E and carotene. Lancet. 337: 1-5.

Rimbach, G., J. Fuchs, and L. Packer (2005) Application of nutrigenomics tools to analyze the role of oxidants and antioxidants in gene expression. pp. 1–12. In: Rimbach, G., J. Fuchs, and L. Packer (Eds.). Nutrigenomics. Taylor and Francis, Boca Raton, FL.

Rodney, J. G. (2004) Antioxidant activity of peanut plant tissues. Master Thesis. Graduate Faculty of North Carolina State University, Department of Food Science. 91 pp.

Rodriguez-Amaya, D. B (2001) A guide to carotenoid analysis in foods. International Life Sciences Institute (ILSI) Press, N.W. Washington, D. C, USA. pp. 71.

Rosegrant, M. W., S. A. Cline, L. Weibo, A. S. Timothy, and A. V.-S. Rowena (2005) Looking ahead long-term prospects for Africa's agricultural development and food security. IFPRI, Washington, D.C. 75 pp.

Sanchez-Moreno, C., J. A. Larrauri, and F. Saura-Calixto (1998) New parameter for evaluation of free radical scavenging capacity of polyphenols. 2nd International Electronic Conference on Synthetic Organic Chemistry, September 1-30. Retrieved May 21, 2008, from: http://www.mdpi.org/escoc/

Scalbert, A., C. Manach, C. Morand, C. Remesy, and L. Jimenez (2005) "Dietary Polyphenols and the Prevention of Diseases. Crit. Rev. Food Sci. Nutr. 45: 287-306.

Schiedt, K. and S. Liaaen-Jensen (1995) Isolation and analysis. Pp. 81-108. In: Britton G., S. Liaaen-Jensen, and H. Pfander (eds). Carotenoids: isolation and analysis, vol 1A. Birkhäuser Verlag, Basel.

Schulzová, V., J. Hajšlová, J. Diviš, and P. Botek. (2006) Organic versus conventional potatoes – is there basis for consumers preference? Paper presented at Joint Organic Congress, Odense, Denmark, May 30-31, 2006.

Schwarz, K., G. Bertelsen, L. R. Nissen,P. T. Gardner, M. L. Heinonen, A. Hopia, T. Huynh-Ba, P. Lambelet, D. McPhail, L. H. Skibsted, and L. Tijburg (2001) Investigation of plant extracts for the protection of processed foods against lipid oxidation. Comparison of antioxidant assays based on radical Lipid-Soluble Antioxidants: Biochemistry and Clinical Applications. Birkhluser, Basel.

Steinmetz, K. A. and J. D. Potter (1996) Vegetables, fruit and cancer prevention: a review. J. Am. Diet. Assoc. 96: 1027 – 1039.

Stock J. S. (1939) A method of graphic interpolation. J. Am. Stat. Ass. 34 (108): 709-713.

Suda, I., S. Furuta, Y. Nishiba, O. Yamakawa, K. Matsugano, and K. Sugita (1997) Reduction of liver injury induced by carbon tetrachloride in rats administered purple-colored sweet potato juice. J. Jap. Soc. Food Sc. Technol. 44: 315-318.

Teow, C. C., V.-D. Truong, R. F. McFeeters, R. L. Thompson, K. V. Pecota, G. C.Yencho (2007) Antioxidant activities, phenolic, and β-carotene contents of sweet potato genotypes with varying flesh colors. Food Chem. 103: 829–838.

Terão, J. (1989) Antioxidant activity of ß-carotene-related carotenoids in solution. Lipids. 24:659-661.

Tonucci, L. H., J. M. Holden, G. R. Beechert, F. Khachik, C. S. Davis, and G. Mulokozi (1995) Carotenoid content of thermally processed tomato-based food products. J. Agric. food Chem. 43: 579-586.

Tsai, P. J., L. Delva, T. Y. Yu, Y. T. Huang, and L. Dufossé (2005) Effect of sucrose on the anthocyanin and antioxidant capacity of mulberry extract during high temperature heating. Food Res. Int. 38: 1059–1065.

Turkmen, N., F. Sari, and S. Velioglu (2005) The effect of cooking methods on total phenolics and antioxidant activity of selected green vegetables. Food Chem. 93: 713-718.

UNICEF (2006) UNICEF analysis of the number of underweight children in the developing world. UNICEF, NY, USA. 36 pp.

United Kingdom Office of Public Sector Information (1990) Food Safety Act 1990 (c. 16). (1990). 11 pp. Retrieved October 21, 2006, from: http://www.opsi.gov.uk/acts/acts1990/ ukpga_ 19900016.

USDA (1975) Composition of foods, USDA, Washington, D.C. 49 pp.

van Poppel, G. (1993) Carotenoids and cancer: an update with emphasis on human intervention studies. Eur. J. Cancer. 29A: 1335 – 1344.

Vinson, J.A., Y. Hao, X. Su, and L. Zubik. 1998. Phenol Antioxidant Quantity and Quality in Foods: Vegetables. J. Agric. Food Chem. 46: 3630-3634.

von Elbe J. H. and S. J. Schwartz (1996) Colorants. pp. 661-672. In: Fennema O. R. (ed.). Food chemistry (3rd edn.). Marcel Dekker, Inc., NY, USA.

Walkowiak-Tomczak, D. (2007) Changes in Antioxidant Activity of Black Chokeberry Juice Concentrate Solutions During Storage. Acta Sci. Pol., Technol. Aliment. 6(2): 49-55.

Walter, W.M. and W.E. Schadel (1981) Distribution of phenols in 'Jewel' sweet potato [*Ipomoea batatas* (L.) Lam.] roots. J Agric Food Chem. 29: 904-906.

Wanda, W. C. (1995) Sweet potato. Retrieved Februar 13, 2007, from the World Wide Web: http://www.hort.purdue.edu/newcrop/ crop fact sheets/ sweetpotato.html.

Wang, C. C., C. Y. Chu, K. O. Chu, K. W. Choy, K. S. Khaw, M. S. Rogers, and C. P. Pang (2004) Trolox-equivalent antioxidant capacity assay versus oxygen radical absorbance capacity assay in plasma. Clin. Chem. 50:952-954.

Wayner, D. D. M., G.W. Burton, K. U. Ingold, and S. Locke (1985) Quantitative measurement of the total peroxyl radical-trapping antioxidant capability of human blood plasma by controlled lipid peroxidation. FEBS Letters. 187: 33-37.

Westman, E. C. (2002) Is dietary carbohydrate essential for human nutrition? Am. J. Clin. Nutr. 75 (5): 951-953.

WFP (2006) Breaking out of the poverty trap. How do we use food aid. Retrieved October 22, 2006 from: http://en. wikipedia. org/wiki/Food#

Whitaker, J. R. (1996) Enzymes. pp. 431-530. In: Fennema O. R. (ed). Food Chemistry (3rd edn.).

WHO/UNICEF (1995) Global prevalence of vitamin A deficiency. WHO micronutrient deficiency information system (MDIS), Working Paper no 2, WHO/NUT/95.3. WHO, Geneva. 116 pp.

Wilhelmina, K., C. F. Forney, A. Martin, and R. L. Prior (1999) Antioxidant capacity, vitamin C, phenolics, and anthocyanins fresh storage of small fruits. J. Agric. Food Chem. 47: 4638-4644.

William, P. J. and A. K. Douglas (2006) Extraction of plant secondary metabolites. pp. 323-351. In: Satyajit DS, Zahid L, Alexander IG (eds). Natural product isolation (2nd edn.). Humana Press, NJ, USA.

Woolfe, J. (1993) Sweet potato: An untapped food resource. Cambridge University Press, Cambridge, USA. 179 pp.

World Bank (1997) World Development Report 1997. The state in a changing world. Oxford University Press for the World Bank, NY, USA. 18 pp. Retrieved October 21, 2006, from http://www.worldbank. org/html.

World Bank (2004) World development indicators 2004. Washington, D.C., USA. 400 pp.

Yagasaki, K., Y. Miura, R. Okauchi, and T. Furuse (2000) Inhibitory Effects of Chlorogenic Acid and its Related Compounds on the Invasion of Hepatoma Cells in Culture. Cytotechnology 33: 229-235.

Yamakawa, O. and M. Yoshimoto (2002) Sweet Potato as Food Material with Physiological Functions. Acta Hort. 583: 179-184 Yap, S.C. (1995) Palm oil carotenes: Chemistry and technology. Ph.D. Thesis, University of Malaya, Kuala Lumpur. 94 pp.

Index

Table I-1. List of abbreviations used in this dissertation

Abbreviation	Definition
ABAP	2, 2'-Azobis(2-amidopropane) hydrochloride
ABTS	2, 2'-Azino-di-[3-ethylbenzthiazoline-6-sulfonate]
AHPA	Acid-hydrolysable phenolic acids
AIDS	Acquired Immune Deficiency Syndrome
AOA	Antioxidant activity
AOP	Anti-oxidative potency in linoleic acid system model
BHA	Butylated HydroxyAnisole
BHPA	Base-hydrolysable phenolic acids
BHT	Butylated HydroxyToluene
B-PE	ß-phycoerythrin
BSA	Bovine serum albumin
CDHP	Conjugated Diene HydroPeroxide
CGIAR	Consultative group on international agricultural research
CIP	Centro international de potato (in Spanish)
d	Day
DE	Diethyl ether
DMPD	4,5-DiMethyl-o-PhenileneDiamine
DPPH	1, 1-diphenyl-2-picrylhydrazyl
EA	Ethyl acetate
EDTA	Ethylenediaminetetraacetic acid
FAO	Food and Agriculture Organization
FPA	Free phenolic acids
FRAP	Ferric Reducing Ability of Plasma or Plants
FRSA	Free radical scavenging activity
GAE	Gallic acid equivalent in mg/g dry extract

Table I-1 (cont.). List of abbreviations used in this dissertation

Abbreviation	Definition
GC-MS	Mass spectrometry with capillary gas chromatography
HAA	Hydrophilic Antioxidant Activity
HPLC	High performance liquid chromatography
HRP	HorseRadish Peroxidase
IPGRI	International food policy research institute
l	Liter
LLP	Liquid-liquid partition
LAA	Lipophilic Antioxidant Activity
LC-MS	Mass spectrometry with liquid chromatography
M	Molar
MDA	3,4-Methyl-2, 5-dimethoxyamphetamine
MGC	MonoGlyceride Citrate
ml	Milliliter
mM	Millimolar
nM	Nanomolar
OD	Oven-drying
PG	n-Propyl Gallate
PHM	pH modification
ROS	Reactive oxygen species
rpm	Revolutions per minute
RT	Room temperature
RVU	Rapid visco unit
SOD	superoxide dismutase
TBA	Thiobarbituric Acid
TBARS	ThioBarbituric Acid Reactive Substances
TBHQ	Tert-ButylHydroQuinone
TCA	TriChloroacetic Acid
TEAC	Trolox equivalent antioxidant capacity

Table I-1 (cont.). List of abbreviations used in this dissertation

Abbreviation	Definition
TEP	Tetraetoxypropane
TNG	Tainong
TPE	Tris-phosphate-EDTA
TPTZ	2,4,6-tripyridyl-s-triazine
Trolox	6-Hydroxy-2,5,7,8-tetramethylchroman-2-carboxylic acid
UV	Ultraviolet
ml	Milliliter
mM	Milimolar
μl	Microliter
μM	Micromolar

Appendix

This appendix section presents supplementary information that are related to, but not essential to understand this dissertation. They organized into Appendix 1, which comprises only Figures and Appendix 2, comprising only Tables.

Appendix 1. Figures

The color parameters in *L, a, b* scale of CIELab (2003) are geometrically illustrated in Figure A-1. Figure A-2 shows a drying curve that was established to determine the end-point drying time of the sweet potato samples at 25, 50, and 75°C. For drying at 25°C, a close-type dryer adjustable between 0 and 50°C was used. For drying at 50 and 75°C, an open-type hot air dryer was used. The drying curve stabilized around a water activity of 0.2-0.4. The drying process took 24, 96, and 240 h to reach this water activity at 25, 50, and 75°C respectively. The freeze-drying process lasted 48 h, while the sun-drying took 7 days (from 10:00 AM to 3:00 PM). Figures A-3 and A-4 show a gallic acid and a catechin calibration curves used to compute the total phenolic and total flavonoid values. Within the concentration range of 0.010 to 0.400 mg gallic acid/ml, the relationship between concentration of gallic acid and absorbance at 765 nm was linear. Within a concentration range of 10 to 500 mg catechin/l, the relationship between concentration of catechin and the absorbance of the sample at 510 nm was linear.

Figure A-5 illustrates the structure of catechin, rutin, and gallic acid. Figure A-6 summarizes the strategies preconized by Arnao *et al.* (1999) for determination of total AOA in chemical compounds and biological samples. A theoretical viscogram of a pasting profile analysis is shown in Figure A-7. Figure A-8 presents the delimitation between skin and flesh on a cross section of TNG73. Figures A-9 and A-10 show products made by filling conventional and Hualien cakes with mashed TNG57 sweet potato.

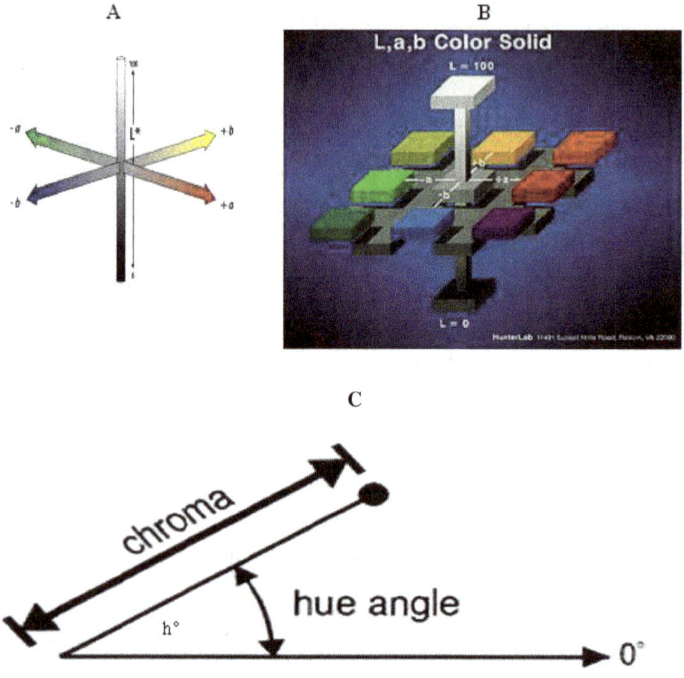

Figure A-1. Geometric illustration of the color parameters: L, a, b, hue angle, and chroma-value (CIELab, 2003).

A: tri-coordinates model showing L, a, and b parameters; B: colorful model showing relationship between L parameter, chroma value and the colors in the visible spectrum region; C: di-coordinates model showing chroma-value and hue angle.

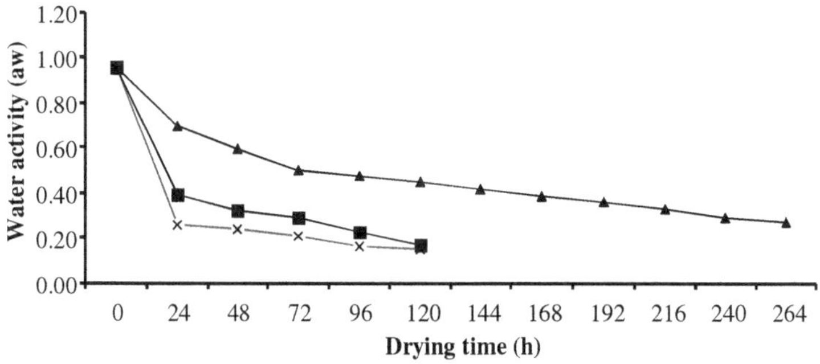

Figure A-2. Drying curve of flesh sweet potato at 25 (▲), 50 (■), and 75°C (x).

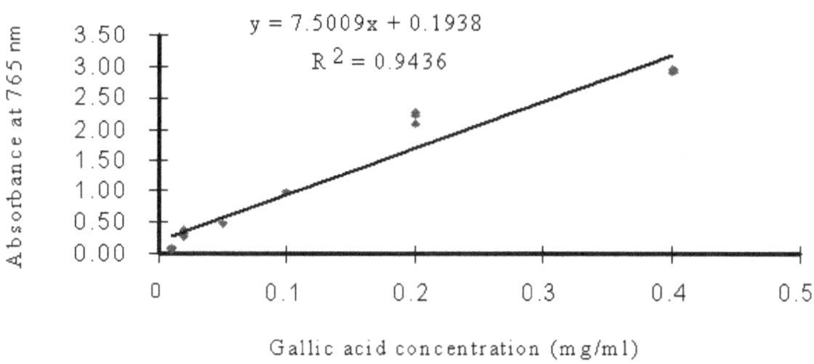

Figure A-3. Standard gallic acid calibration curve for total phenolic assay.

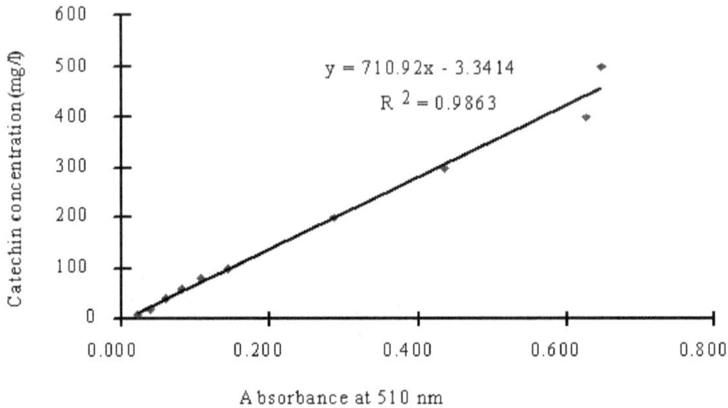

Figure A-4. Standard catechin calibration curve with catechin concentration in Y-axis for total flavonoids assay.

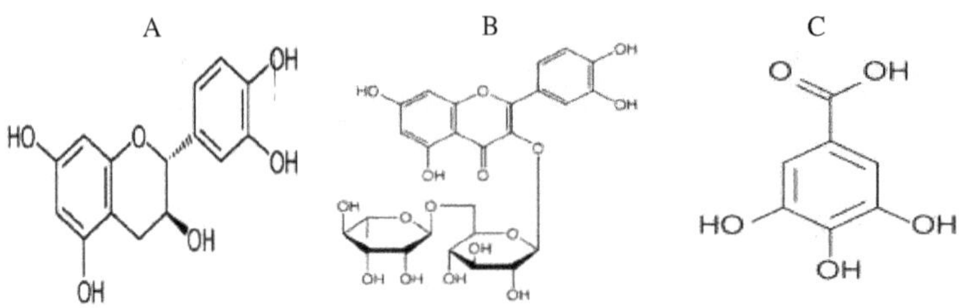

Figure A-5. Structure of catechin (A), rutin (B), and gallic acid (C).

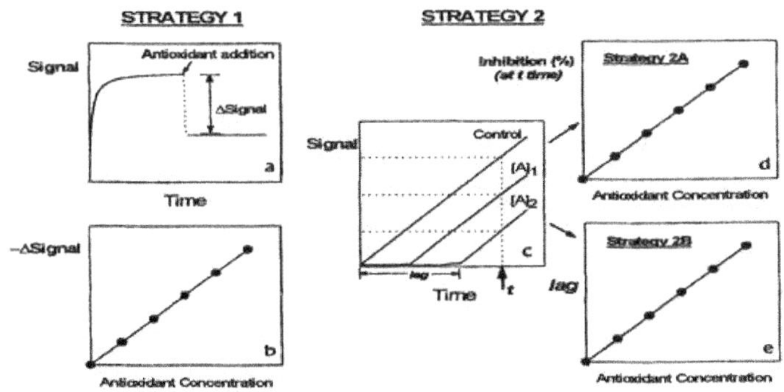

Figure A-6. Strategies for determination of total AOA in chemical compounds or biological samples. Source: Arnao et al. (1999).

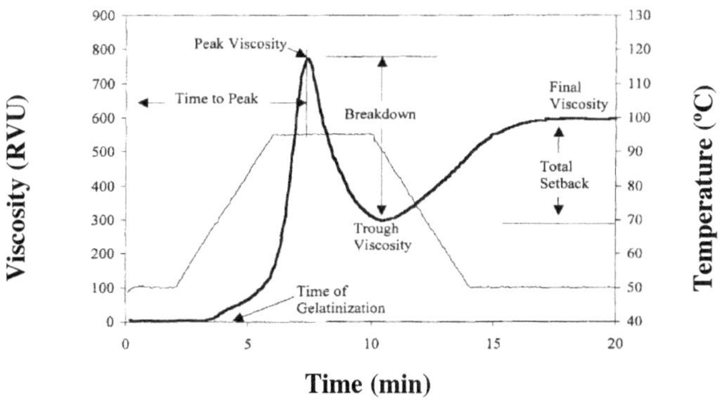

Figure A-7. A theoretical viscogram showing RVA viscosity profile (bold line) and temperature profile (normal line) depicting time to gelatinization, time to peak, peak viscosity, breakdown, trough viscosity, total setback, and final viscosity (Higley *et al.*, 2003).

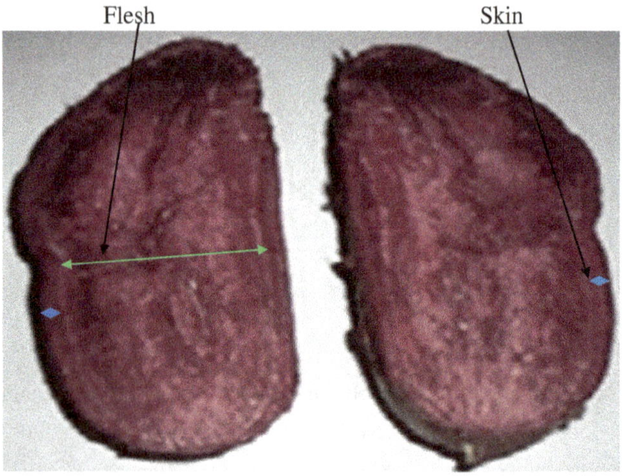

Figure A-8. Identification of skin and flesh of TNG73 sweet potato. The central part indicated by the green double arrow is the flesh; the outer part indicated by the blue double arrow is the skin.

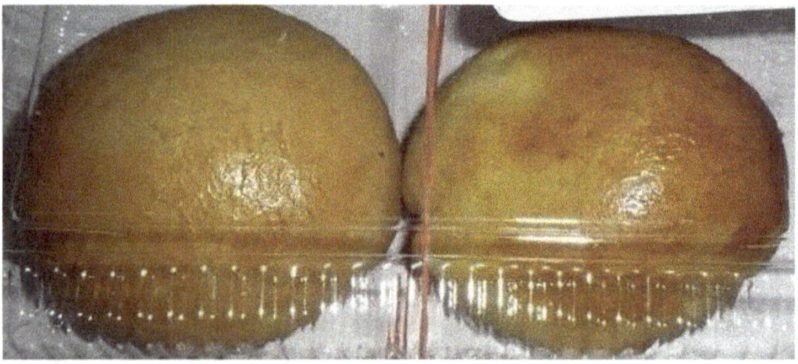

Figure A-9. Hualien cake filled with mashed TNG57. The crust is made of wheat flour; the interior is filled with sweet potato.

A C

B D

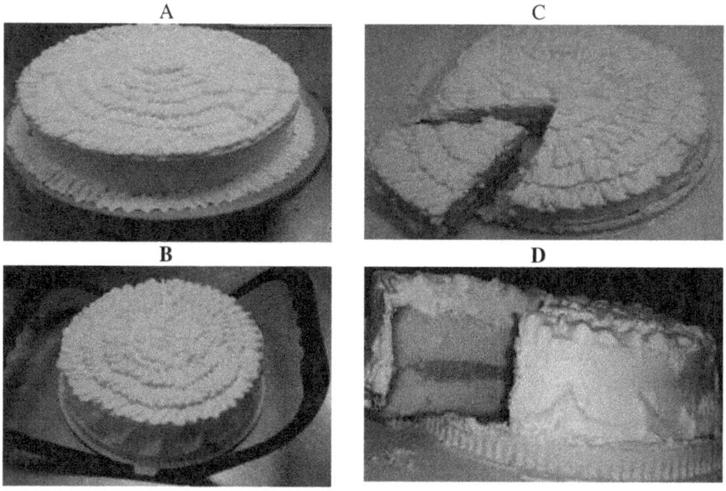

Figure A-10. Conventional cake filled with mashed TNG57. A: the cake just finished decoration. B: the on a plate pending covering;. C and D: two views of the cake being cutting for eating; the filling mashed sweet pototo can be seen on the profile.

Appendix 2. Tables

The factors and factor levels evaluated in this dissertation are presented in Table A-1. Table A-2 shows the response variables influenced by these factors. The data in Table A-3 were generated from Figure A-1 to interpret the hue angle (h°) values.

The basic colors red, yellow, green, and blue, correspond to the h° values of 0, 90, 180, and 270°, respectively (Figure A-1, part B). However, a color corresponds to a range of hue values, and two basic colors are separated by intermediate colors. For example, yellowish red and reddish yellow separate red from yellow, while greenish yellow and yellowish green separate yellow and green. Similarly, Bluish green and greenish blue are between green and blue, while reddish blue and bluish red separate blue from red. Therefore, we divided the hue difference of 90° between two closest basic colors by 4, which gave a value of 22.5°. Consequently, red corresponded to h°=0±22.5°, while yellow corresponded to h°=90±22.5°. Similarly, green corresponded to 180±22.5°, while blue corresponded to 270±22.5°. The upper hue limit of a color was considered as the lower hue limit of the next color. For example, 22.5° was the upper limit of red and the lower limit of yellowish red. To determine the upper limit of an intermediate color, we added 22.5° to its lower limit. For example, the upper limit of yellowish red was computed as 22.5°+22.5°=45°. Similarly, to determine the upper limit of reddish yellow, we added 22.5° to its lower limit, which gave 67.5° that was also the lower limit of yellow.

Dry matter contents of samples subjected to different drying methods are shown in Tables A-4 and A-5, while Table A-6 presents the 10 most important sweet potato cultivars in Taiwan. Table A-7 shows average energy and protein productions of selected food crops, and Table A-8 shows the UK recommendations for protein intakes. Table A-9 presents some commonly used organic solvents with their polarity index, boiling point, and density.

Table A-1. Factors and Factors' levels studied in this dissertation

---Factor---		---Treatment---	
No.	Name	No.	Name
1	Sweet potato Genotype	1	TNG57
		2	TNG66
		3	TNG73
2	Part of the root	4	Skin
		5	Flesh
		6	Whole root
3	Extracting solvent	7	50% aqueous ethanol
		8	Acetone
		9	Diethyl ether
		10	Distilled de-ionized water
		11	Ethanol 95%
		12	Ethyl acetate
		13	Methanol
4	Drying method	14	Freeze-drying
		15	Low-temperature-drying at 25°C
		16	Hot air-drying at 50°C
		17	Hot air-drying at 75°C
		18	Sun-drying
		19	Non-drying
5	Fractionation method	20	Liquid-liquid partition
		21	pH modification
6	Fraction	22	Acid hydrolysable phenolic fraction
		23	Base hydrolysable phenolic fraction
		24	Free hydrolysable phenolic fraction
		25	Diethyl ether fraction
		26	Ethyl acetate fraction

Table A-1 (cont.). Factors and Factors' levels studied in this dissertation

---Factor---		---Treatment---	
7	Storage temperature	27	Room temperature
		28	4°C
8	Storage duration	40	Months 1 to 12

Table A-2. Response variables studied in this dissertation

---Complex variable---		---Single variable---	
No.	Name	No.	Name
1	Color parameters	1	Color lightness
		2	Hue angle
		3	Chroma
2	Antioxidant activity	4	DPPH free radical scavenging activity
		5	ABTS free radical scavenging activity
		6	Antioxidative potency in linoleic acid system
		7	DPPH IC50
		8	ABTS IC50
		9	Relative desirability index (RDI)
3	Phenolic content	10	Total phenolics
		11	Total flavonoids
		12	Catechin
		13	Gallic acid
		14	Rutin
4	Carotenoid content	15	Total carotenoids
		16	Beta-carotene

Table A-3. Interpretation keys of the hue angle (h°) values corresponding to the visible color of the samples analyzed

Hue angle (h°)	Corresponding visible color
$0.0 < h^{o} \leq 22.5$	Red
$22.5 < h^{o} \leq 45.0$	Yellowish red
$45.0 < h^{o} \leq 67.5$	Reddish yellow
$67.5 < h^{o} \leq 112.5$	Yellow
$112.5 < h^{o} \leq 135.0$	Greenish yellow
$135.0 < h^{o} \leq 157.5$	Yellowish green
$157.5 < h^{o} \leq 202.5$	Green
$202.5 < h^{o} \leq 225.0$	Bluish Green
$225.0 < h^{o} \leq 247.5$	Greenish blue
$247.5 < h^{o} \leq 292.5$	Blue
$292.5 < h^{o} \leq 315.0$	Reddish blue
$315.0 < h^{o} \leq 337.5$	Bluish red
$337.5 < h^{o} \leq 360.0$	Red

Table A-4. Dry matter content (%) of the flesh of TNG57 and TNG66 sweet potato samples submitted to different drying conditions

Drying conditions[2]	Dry matter content (mean±SE, n=3)[1]
TNG57	
Freeze-drying, -40°C	97.35±0.35a
Sun-drying	90.63±0.28b
LTD, 25°C	88.72±0.31b
HAD, 50°C	90.06±0.29b
HAD, 75°C	95.55±0.20a
TNG66	
Freeze-drying, -40°C	94.08±0.02a
Sun-drying	91.30±0.25b
LTD, 25°C	91.23±0.35b
HAD, 50°C	95.51±0.28b
HAD, 75°C	94.71±0.20a

[1]Means with the same letter in a column are not significantly different ($p>0.05$); [2]LTD: low-temperature drying; HAD: hot air drying.

Table A-5. Dry matter content (%) of TNG73 samples submitted to different drying conditions

Drying conditions[2]	Dry matter content (mean±SE, n=3)[1]
Flesh	
Freeze-drying, -40°C	98.23±0.05[a]
Sun-drying	93.54±0.30[b]
LTD, 25°C	88.33±0.29[c]
HAD, 50°C	92.07±0.31[b]
HAD, 75°C	93.24±0.18[b]
Skin	
Freeze-drying, -40°C	97.45±0.30[a]
Sun-drying	93.10±0.09[b]
LTD, 25°C	92.50±0.51[b]
HAD, 50°C	91.99±0.27[b]
HAD, 75°C	91.20±0.31[b]

[1]Means with the same letter in a column are not significantly different ($p>0.05$); [2]LTD: low-temperature drying; HAD: hot air drying.

Table A-6. List of the 10 most important sweet potato cultivars and key characteristics being subject to breeding during 1956- 2003 in Taiwan

Cultivar	Characteristic	Remark
TNG57	For steaming, boiling, baking, and roasting; orange flesh, tastes good, generating fast.	Denominated in 1956
Tainong 64	For cooking; orange flesh, tastes good, suitable to be cultivated in fall.	Denominated in 1977
Tainong 65	For cooking and feedstuff; high protein content, suitable to be harvested in fall.	Denominated in 1981
TNG66	For eating; reddish flesh, high carotene content, strong adaptability.	Denominated in 1982
Tainong 67	For eating; orange flesh, resistant to sweet potato weevil, suitable to be cultivated in fall.	Denominated in 1985
Tainong 68	For processing and leafy vegetable in summer; high amylum content, fast growth of stems and leaves.	Denominated in 1985
Tainong 69	For eating; orange flesh, section in brown intensity low.	Denominated in 1991
Tainong 70	For eating; orange flesh, section in brown intensity low.	Denominated in 1993
Tainong 71	For leafy vegetable; fine leaf quality, less bitter, high vitamins A, and C, and folic acid contents, green color after cooking, tastes good, the whole year cultivation.	Denominated in 1998
Tainong 72	For eating and raw material of roast; spindle in tuberous root; cutis in deep red, smooth; orange flesh; harvest in fall; 0.3 ton/acre; about 28-32% dry matter in root tubers; 4.63 mg reducible carbohydrate /100 g fresh root tubers; 3773 carotene unit/100 g fresh root tuber,.	Denominated in 2003

Source: Council of Agriculture of Taiwan (CAO, 2006).

Table A-7. Comparison of average energy and protein production of selected food crops in developing countries

Crop Crop	Comparison parameters[1]				
	A	B	C	D	E
Potato	130	18	54	1.5	12.60
Yam	180	14	47	1.0	8.80
Sweet potato	180	22	70	1.0	6.70
Rice, paddy	145	18	49	0.9	3.40
Groundnut in shell	115	8	36	1.7	2.60
Wheat	115	14	40	1.6	2.30
Lentil	105	6	23	1.6	2.30
Cassava	272	13	27	0.1	2.20

[1]A: Growth duration (days); B: Dry matter (kg/ha/ day); C: Edible energy (1000 kcal/ha/day); D: Edible protein (kg/ha/day); E: Production value (US$/ ha/day). Sources: FAO (1984); USDA (1975).

Table A-8. The UK reference nutrient intakes (RNI) for protein[1]

Type of person	Amounts required (g/day)
Infants/Children	
0-12 months	12.5-14.9
1-3 years	14.50
4-10 years	19.7-28.3
Boys	
11-14 yrs	42.1
15-18 yrs	55.2
Girls	
11-14 yrs	41.2
15-18yrs	45.0
Men	
19-50 yrs	55.5
50+ yrs	53.3
Women	
19-50 yrs	45
50 years and over	46.5
During pregnancy	Extra 6g/day
Breast feeding 0-6months	Extra 11g/day
Breast feeding 6 months and over	Extra 8g/day

[1]The recommendations are based on the complete digestibility of milk or egg protein. Since protein from plant sources may be slightly less digestible, the UK's Department of Health recommends that vegetarians multiply the above figures by a factor of 1.1.

Sources: FAO/WHO/United Nation University (UNU) (1985); Milward *et al.* (1992).

Table A-9. Physical characteristics of some common solvents

Solvent	Polarity index	Boiling point (°C)	Density (g/ml)
Hexane	0.0	69	0.655
Diethyl ether	2.8	35	0.713
Dichloromethane (DCM)	3.1	40	1.326
Isopropanol	3.9	82	0.785
Tetrahydrofuran (THF)	4.0	66	0.886
n-Butanol	4.0	118	0.810
Chloroform ($CHCl_3$)	4.1	61	1.498
Ethyl acetate	4.4	77	0.894
Acetone	5.1	56	0.786
Methanol	5.1	65	0.791
Ethanol	5.2	79	0.789
Acetonitrile (MeCN)	5.8	82	0.786
Acetic acid	6.2	118	1.049
Dimethylformamide (DMF)	6.4	153	0.944
Dimethyl sulfoxide (DMSO)	7.2	189	1.092
Water	9.0	100	1.000

Source: extracted from http://en.wikipedia.org/wiki/Solvent (retrieved 2007, July 19th); Seidel (2006).

Biosketch (Vitae) of the author

Full name: Predner **DUVIVIER**

Gender: Male

Postal term address: 1 Hseuh Fu Rd, Neipu Hsiang, 13 Pigntung 912, DTAIC, **NPUST,** Taiwan.

Postal permanent address: Faculté d'Agronomie et de Médecine Vétérinaire (**FAMV**), Route National No. 1, Damien, Port-Au-Prince, Haiti (WI). BP. 1441.

E-mail address: pduvivier@yahoo.com

Birthday: Haiti, November 27

Education

Primary school: 1976-1982: Ecole Nationale de Javel, Anse- A-Veau, Haiti

Secondary school: 1982-1989: Lycée Alexandre Pétion, Port-au-Prince, Haiti.

Undergraduate university study : 1989-1994: Université d'Etat d'Haiti (**UEH**), Faculté d'Agronomie et de Médecine Vétérinaire (**FAMV**). Diploma: Ingénieur Agronome, option Phytotechnie (Agricultural Engineer, major: Plant Science).

Master's study: 1997-1999: University of Puerto Rico (**UPR**), Mayaguez Campus (**RUM**), College of Agricultural Science. Diploma: M. Sc. in Horticulture, major: Fruit crops.

Doctorate study: 2006-2009: National Pingtung University of Science and Technology (**NPUST**), Department of Tropical Agriculture and International Cooperation (**DTAIC**), Taiwan. Diploma: Ph. D., major: Food Science.

Work experience

1995-1996: Haitian Association for Monitoring Water and Soils (**ASSODLO**). Responsibilities: technical assistance to farmer associations; agricultural extension.

1996-to date: Université d'Etat d'Haiti (**UEH**), Faculté d'Agronomie et de Médecine Vétérinaire (**FAMV**). Responsabilities: teaching, research, and services.

Teaching area: plant propagation, agricultural biometrics, and fruit crops.

Some administrative services

April 2002- August 2006: Chairman of the **FAMV** experimental Research Farm in Damien.

April 2005-August 2006: Chairman of the Department of Phytotechny of **FAMV**.

October 2009-April 2013: Chairman of the department of basic sciences of **FAMV.**

October 2010-April 2013: Chairman of the Laboratory of Chemistry of **FAMV.**

April 2013-present: Vice-Dean for academic affairs of **FAMV**.

March 9th, 2025: Elected Vice-Rector for academic affairs of **UEH** for a 4-year term. (Inauguration to take place on March 19th, 2025; inauguration as Vice-Rector results automatically in stepping down from the position of Vice-Dean).

Periodical publications

Carvil O.N, Tescar R.-P., Doréus G., Duvivier P., Exilien R., Joseph F., Augustin H. 2024. Contrôle de la pourriture brune des cabosses du cacaoyer (*Theobroma cacao* L.) à l'Anse-d'Hainault (Grand'Anse, Haïti). RED 11 (1): 3 – 7.

Duvivier, P., Doréus, G., Tescar, R. P., & Carvil, O. N. 2024. Agronomic characteristics and economic performance of cocoa farms in Haiti. Polish Journal of Agronomy/ Current

Agronomy, (1), 44-53. doi: 10.2478/cag-2024-0005. Received 8 June 2024, reviewed 10 September 2024, accepted 27 September 2024.

DUVIVIER Predner, DUVIVIER Predner Junior, DOREUS Gelin, TESCAR Robers-Pierre et CARVIL Ophny Nicolas. 2024. Corrélation entre les données météorologiques satellitaires et in situ en Haïti: cas de la pluviométrie et de la température. RED 11(1) : 24-27. Reçu 27 juin 2024. Accepté 23 septembre 2024.

DUVIVIER Predner, DOREUS Gelin, TESCAR Robers-Pierre, AUGUSTIN Holken, DUVIVIER Predner Junior et CARVIL Ophny Nicolas. 2024. Rotation culturale et fertilisation minérale améliorent la fertilité des sols et le rendement de la carotte (Daucus carota L.) sur le Plateau des Rochelois, Haïti. RED 11(1) : 8-15. Reçu 12 juin 2024. Accepté 23 septembre 2024.

Romain Exilien, Robers Pierre Tescar, Holken Augustin, Dorensky Cadestin, Nicolas Carvil, Predner Duvivier, Gelin Doreus, Jessi Griesheimer. 2024. Impact of planting and harvest dates on yam infestation by white grubs. Journal of applied entomology, Wiley. DOI: 10.1111/jen.13348. Received: 24 June 2024 | Revised: 21 August 2024 | Accepted: 23 August 2024

Duvivier Predner, Tescar Robers Pierre, Halliday Cassandra, Murphy Madhuvanti C., Guell Cornelia, Howitt, Augustus Eden, Haynes Emily and Unwin Unwin. 2024. Differences in income, farm size and nutritional status between female and male farmers in a region of Haiti, volume 8. Front. Sustain. Food Syst. DOI: 10.3389/fsufs.2024.1275705. Received: 10 Aug 2023; Accepted: 01 Apr 2024.

Leonor Guariguata, Gordon M. Hickey, Madhuvanti M. Murphy, Cornelia Guell, Viliamu Iese, Karyn Morrissey, Predner Duvivier, Stina Herberg, Sashi Kiran, Nigel Unwin. 2023. Understanding the links between human health, ecosystem health, and food systems in Small Island Developing States using stakeholder-informed causal loop diagrams. PLOS Glob Public Health 3(9): e0001988. https://doi.org/10.1371/journal.pgph. 0001988

Halliday C, Morrissey K, Saint Ville A, Guell C, Augustus E, Guariguata L, Iese V, Hickey G, Murphy MM, Haynes E, Tescar RP, Duvivier P and Unwin N. 2023. Trends in food supply, diet, and the risk of non-communicable diseases in three Small Island Developing States: implications for policy and research. Front. Sustain. Food Syst. 7:1058540. doi: 10.3389/fsufs.2023.1058540.

Carvil O. N., Tescar R. P. et Duvivier P. 2023. La maladie du balai de sorcière (*Moniliophthora* perniciosa) sur cacaoyer (*Theobroma cacao* L.) identifiée pour la première fois en Haïti. RED 10 (1): 16 – 19.

Charles L. J., Duvivier P. et Félix B. A. **2019**. Bilan carbone à l'échelle des exploitations agricoles en Haïti : cas du bassin versant de Limbé. RED 8 (2) : 34-40.

Carvill, O. N., Tescar, R. P., Duvivier P. et Jean-Baptiste N. **2016**. Le Système de Riziculture Intensive (SRI) : atouts, contraintes et perspectives d'utilisation en Haïti. RED 7(2): 36-46.

Duvivier P., Joseph D. et Antoine J. S. **2016.** Le coût économique et écologique de la pratique du brûlis dans l'agriculture haïtienne : cas des pertes d'azote et de carbone dans les rizières de la Vallée de l'Artibonite. RED 7(2): 47-50.

Duvivier P. et Orélien F. **2016**. Problématique de l'eau dans le sous bassin versant Ravine Diable (Saut du Baril), troisième section communale de l'Anse-à-Veau, Haïti. RED 7(2): 19 - 25.

Duvivier, P., Antoine J. S., Joseph D., Alphonse M. E. et Jean Baptiste F. **2016**. Efficience d'utilisation de l'azote par trois phénotypes de riz (*Oryza sativa*, L.) sur la Ferme de Mauger, Vallée de l'Artibonite, Haïti. RED 7(2): 3-10.

Duvivier P., Blaise J. et Gonomy N. 2015. Qualités physico-chimiques de l'eau dans le bassin versant d'Ennery-Quinte, Artibonite. RED 7(1): 14 - 23.

Saint-Vil B., Corantin H., Duvivier P. et Val M.E. 2014. Mise au point d'un dispositif familial de traitement de l'eau de boisson par filtration sur sable suivie de désinfection solaire. RED 6 (1):16 - 21.

Duvivier P., Antoine J. S., Joseph D. et Alphonse M. E. 2014. Mesure du rendement et de la biomasse de trois phénotypes de riz (*Oriza sativa* L) en réponse à quatre doses d'azote sur la ferme de Mauger, Vallée de l'Artibonite. RED 6 (1): 10- 15.

Duvivier P., Hsieh P.-C., Lai P.-Y. and Charles A. L. 2012. Evolution of the total phenolic content and antioxidant activity of Tainong 73 (TNG73) sweet potato (*Ipomoeabatatas* Lam) flour during a 12-month storage period. RED 5 (1): 915.

Duvivier P., Hsieh P.-C., Lai P.-Y. and Charles A. L. 2010. Retention of phenolics, carotenoids, and antioxidant activity in the Taiwanese sweet potato (Ipomoea batatas Lam.) CV Tainong 66 subjected to different drying conditions. African Journal of Food, Agriculture, Nutrition and Development (AJFAND) 10 (11):4414-4431.

Duvivier P., P.-C. Hsieh, P.-Y. Lai, and A. L. Charles (2009) Effect of freeze-drying on phenolic content and antioxidant activity of Tainong 57, Tainong 66, and Tainong 73 cultivars of sweet potato (*Ipomoea batatas* Lam.). J. International Cooperation (JIC) 4(1): 1-15.

Duvivier P., P.-C. Hsieh, P.-Y. Lai, and A. L. Charles (2008) Total phenolic content and antioxidant activity of seven extracts of sweet Potato (*Ipomoea batatas* Lam.) as affected by the extracting solvent and the genotype. Revue Scientifique de la FAMV: Recherche, Etude, Développement (RED) 4 (1): 20-27.

Duvivier P., P.-C. Hsieh, P.-Y. Lai, and A. L. Charles (2008) Effects of the drying process on the antioxidant activity, phenolic, and carotenoid contents of sweet potato (*Ipomoea batatas* Lam.) variety Tainong 57. Revue Scientifique de la FAMV: Recherche, Etude, Développement (RED) 4 (1): 43-48.

Duvivier P., P.-C. Hsieh, P.-Y. Lai, and A. L. Charles (2008) Evaluation of drying methods on Antioxidant Activity, Total Phenolic and Total Carotenoid

Contents of Sweet Potato (*Ipomoea batatas* (L.) Lam.) var. Tainong 73. J. International Cooperation (JIC) 3(1): 73-86.

Duvivier P., J. Louissaint, et U. Sampeur (2006) Réponse de trois variétés de riz (*Oriza sativa* L) [Bogapoté, Malaika, et TCS10] à la fertilisation phosphatée et potassique dans la Vallée de l'Artibonite, Haïti. *Revue* Scientifique de la FAMV: Recherche, Etude, Développement (RED) 3 (1):22-25.

Duvivier P., D. J. Eyma, M. L. Fontin, R.-P. Tescar, et E. J. Vilmont (2005) Vers la réhabilitation durable de la ferme expérimentale de Damien. Revue Scientifique de la FAMV: Recherche, Etude, Développement (RED) 2 (1):51-52.

Duvivier P. et F. Moise (2005) Effets de deux anti-gibbérellines (Bonzi et B-nine sp) sur la croissance de deux variétés (Roma VF et Ace 55 VF) de tomate (*Lycopersicum esculentum* Mill) à Fermath, Haïti. Revue Scientifique de la FAMV: Recherche, Etude, Développement (RED) 2 (1):15-17.

Louissaint J. et P. Duvivier (2005) Principales caractéristiques des terres rizicoles de la Vallée de l'Artibonite (Haïti) et disponibilités réelles en éléments majeurs. Revue Scientifique de la FAMV : Recherche, Etude, Développement (RED) 2 (1):32-36.

Duvivier P. et C.-A. Maldonado (2003) Evaluation de cinq porte-greffes de manguier (*Mangifera indica* L) pour l'efficience de production des variétés commerciales "Parvin" et Tommy Atikins". Revue Scientifique de la FAMV: Recherche, Etude, Développement (RED) 1 (1):15-22.

Conference presentation

Duvivier P., P.-C. Hsieh, P.-Y. Lai, and A. L. Charles (2008) Comparative evolution of antioxidant activity, total phenolic and carotenoid contents during the drying process in three varieties of sweet potato (*Ipomoea batatas* Lam.) cultivated in Taiwan. P.740-750. In: Proceedings of the 5th Taiwan-

Thailand Bilateral Conference on International Collaboration on Tropical Agriculture: Agriculture for Improving Human Life. May 7-9, 2008. Pingtung, Taiwan. Published by National Pingtung University of Science and Technology (NPUST), Taiwan.

Duvivier P. and Maldonado C.-A. 2000. Evaluation of mango (*Mangifera indica* L.) rootstocks for yield efficiency of 'Parvin' and 'Tommy Atkins' varieties. Proceedings of the Caribbean Food Crops Society held in St Lucia 1999. 35: 144-154.

Duvivier P., Maldonado C.-A y Alvaro Acosta. 2000. El Patron de of mango (*Mangifera indica* L.) y sus efectos sobre la caida prematura de las frutas. Sociedad Caribena de cultivos alimenticios, 36 memoria annual, Boca Chica, Republica Dominicana 2000. 36: 20-26.

Working paper

Duvivier P. et Fontin Marie Lesly. 2017. Building the Evidence Base on the Agricultural Nutrition Nexus: Haiti. CTA Working Paper 17/09| June 2017. Series: Agriculture and nutrition. State University of Haiti, Apia. ISBN 978 982 9003 88 1.

Memoire, thesis, and dissertation

Duvivier P. (1995) Caractérisatisation de huit populations de haricot (*Phaseolus vulgaris* L) noir originaire du Centre de Salagnac, Haïti. [Mémoire d'Ingénieur Agronome]. Faculté d'Agronomie et de Médecine Vétérinaire (FAMV), Université d'Etat d'Haiti (UEH), Port-au-Prince, Haïti.

Duvivier P. (1999) Caracterización de dos variedades comerciales de mango (*Mangifera indica* L.) 'Parvin' y 'Tomy Atkins' con cinco patrones diferentes à Fortuna, Puerto Rico. [Tesis de Maestría]. Recinto Universitario de Mayagüez (RUM), Universidad de Puerto Rico (UPR), Mayagüez, Puerto Rico (PR).

Duvivier P. (2009) Antioxidant activity of sweet potato (*Ipomoea batatas* (L.) Lam.). [Ph. D. Dissertation]. National Pingtung University of Science and Technology (NPUST), Department of Tropical Agriculture and International Cooperation (DTAIC), Pingtung, Taiwan.

My biggest dream

My biggest dream is a world where at least the basic human rights are achieved, particularly the right to eat, to live, and to live in dignity.

My biggest goal

My biggest goal is to contribute to improve the food and nutrition security in my country; to promote healthier nutrition status and lifestyle to my people; to contribute to the education of the yough because educated people do not accept to live and die in poverty, but they seek ways to fight poverty and create wealth.

Perspectives

To date, I have always worked in the area of food production. With the new knowledge and tools adquired during my Ph. D. study, I am looking at food science and technology as a bridge to build between harvest seasons usually characterized by food surpluses and inter-harvest seasons that often correspond to food shortage. The bridge is also to be built between production and comsumption areas. Beside food production, food processing, food preservation and storage, and food distribution will be given high priority.

www.ingramcontent.com/pod-product-compliance
Lightning Source LLC
LaVergne TN
LVHW052259200926
848481LV00002B/409
9798892489188